MW01538397

SIMULATION OF UNDERGROUND GAS STORAGE IN A DEPLETED FIELD.

Donald Kuiekem

SIMULATION OF UNDERGROUND GAS STORAGE IN A DEPLETED FIELD.

This work promotes the rehabilitation of depleted deposits into underground storage sites.

ScienciaScripts

Imprint

Any brand names and product names mentioned in this book are subject to trademark, brand or patent protection and are trademarks or registered trademarks of their respective holders. The use of brand names, product names, common names, trade names, product descriptions etc. even without a particular marking in this work is in no way to be construed to mean that such names may be regarded as unrestricted in respect of trademark and brand protection legislation and could thus be used by anyone.

Cover image: www.ingimage.com

This book is a translation from the original published under ISBN 978-620-3-41545-2.

Publisher:
Sciencia Scripts
is a trademark of
Dodo Books Indian Ocean Ltd., member of the OmniScriptum S.R.L Publishing group
str. A.Russo 15, of. 61, Chisinau-2068, Republic of Moldova Europe
Printed at: see last page
ISBN: 978-620-3-63416-7

To my dear mother who has always supported and
encouraged me throughout my life.

MANYI MONICA KAMGANG

ACKNOWLEDGEMENTS

At the end of this work, I would like to express my deepest thanks to all those who have supported me in any way and have always been there for me.

First of all, I thank God, the Almighty for all his blessings and mercy that he never ceases to grant me.

I would like to thank Prof. Abdoul WAHABOU, Director of the School of Geology and Mining of the University of Ngaoundéré for allowing us to defend this Master's thesis.

I also thank Dr. Madeleine TCHUENTE, Minister of Scientific Research and Innovation for her support to the youth willing to do research and for all the efforts made to develop and promote scientific research in Cameroon.

I then thank Professor Ismaïla Ngounouno, Director of the School of Geology and Mining of the University of Ngaoundéré from 2011 to 2019 for this quality training.

I would like to thank my supervisor Prof. Arsène MEYING, Head of the Division of Schooling, Studies and Training, for his availability, his expertise, his patience, his advice and his assistance during the completion of this work.

I would like to thank my supervisor Dr. Ing Luc Leroy MAMBOU NGUEYEP, Head of the Oil and Gas Production Department, for his availability, his advice, and his perpetual moral and material assistance, which he gave me for the completion of this work.

I thank my technical supervisors Mr. Emmanuel ATANGANA, PhD student at the Technical University of Freiberg and Mr. Victorien DJOTSA, Researcher and Assistant Professor at the University of Kiel in Germany.

With the same impetus, I thank all the teachers of the EGEM who have given the quality teachings that make up this training and all the administrative and support staff.

I would like to thank my parents Mr. and Mrs. KUIEKEM, my tutor Mrs. MOUATCHUI Cécile and the MOXEBO family as well as my brothers and sisters Joel KUIEKEM, Grace KUEKEM, Gilles KUIEKEM and Annita NGANGOUE for their support.

I thank then my friends Marcel EKOO, Ivan SORE, Jean AWA AWA, Kelly SIDIANG ELOGO, Kevin ELOGO, Valérie BENDEGUE, William BOROH, Cindy KADJE, Cyrille KAMDEM, Dimitri DJEUKOUA, Vittel FOSSO, Joséphine MATATEYOU, Franklin GUELOWHEU, MONEGNOKO BABOGA, TEMGOUA Stella, ABENDE SAYOM, Annie NOUKI and especially Francesca KIYOBO for their assistance and affection.

Finally my general thanks go to :

- all my family who continue to be there for me;

- all my fellow students of the research Master.

Those whose names are not listed here should not feel forgotten.

Summary

In a world in full energy transition, the developed countries accumulate strategic reserves in order to periodically compensate for a possible crisis thanks to the methods of underground storage of energy resources. The reason why we have articulated this work around the simulation of underground gas storage in a depleted reservoir in order to highlight the main mechanisms of flow and conservation of gas in a depleted reservoir and to determine the conditions for an optimal storage.

To achieve this work, we used the CMG software (Computer Modelling Group) which allowed us to design the 3D model of an oil reservoir with a normal fault. We simulated the stages of the life of this field, namely the primary and secondary recovery, and we obtained an average recovery of 65% of hydrocarbons. Different gas injection and production scenarios over a 50-year period were simulated in order to analyze the behavior of fluids at depth and to determine the conditions required to efficiently store gas in a depleted field. In the course of this work, we were able to highlight the barrier phenomenon represented by a normal fluid flow fault and observe the losses during gas production due to its change of state.

Keywords: Strategic reserves, underground storage, depleted field, simulation, gas, primary and secondary recovery.

Abstract

In a world in the midst of energy transition, developed countries are accumulating strategic reserves in order to periodically compensate for a possible crisis by using underground storage methods of energy resources. The reason why we have articulated this work around the simulation of the underground gas storage in a depleted deposit in order to highlight the main mechanisms of gas flow and conservation in an exhausted reservoir and to determine the conditions for an optimal storage.

To carry out this work, we used the CMG (Computer Modeling Group) software which allowed us to design the 3D model of an oil reservoir with a normal fault. We simulated the stages of life of this field, namely the primary and secondary recovery, then we obtained an average recovery of 65% of hydrocarbons. Various gas injection and production scenarios over a 50-year period were simulated to analyze the behavior of fluids at depth and determine the conditions required to efficiently store gas in a depleted field. During this work, we were able to highlight the barrier phenomenon represented by a normal fault to the flow of fluids and observe the losses during gas production due to its change of state.

Keywords: Strategic reserves, underground storage, depleted deposit, simulation, gas, primary and secondary recovery.

Table of contents

List of Figures

List of tables

List of Abbreviations

IEA: International Energy Agency

B_g : Formation gas volume factor

B_o : Formation oil volume factor

B_w : Volume of formation factor

CO_2 : Carbon dioxide

E_f : Expansivity of the rock (formation)

E_g : Gas expansion

E_o: Oil Expansivity

E_w : Water Expansivity

LPG : Liquefied Petroleum Gas

OECD: The Organization for Economic Cooperation and Development

OOIP: "Original Oil In Place

OPEC: Organization of the Petroleum Exporting Countries

PVT: Pressure, Volume and Temperature

R_s : Ratio of a gas solution in oil phase

R_v : Ratio of vaporized oil in gas phase

SCDP : Cameroonian Oil Deposits Company

S_g : Gas saturation

S_o : Oil saturation

S_w : Water saturation

USA: "United States of America

Φ: Porosity

General introduction

Storing oil is an economic and strategic imperative. It is impossible to imagine a chain running from the wellhead in the producing fields to the consumer's gas pump without storage. Demand fluctuations are numerous and lead to consumption peaks: daily, especially for domestic fuels (gas, heating oil); weekly, for automotive fuels; seasonal for heating fuels (gas, liquefied gas and heating oil). Production, transport (pipelines, gas pipelines, tankers) and refining facilities capable of absorbing all these peaks would be uneconomic, hence the need to distribute regulatory capacity: on the oil fields, to maintain a shipping buffer, ensure collection from wells and compensate for production incidents; at the ends of the pipelines : because tanker offtake or arrivals, with loading and unloading rates much higher than those of the pipelines, require significant storage, which is also essential for the segregation of crude oils of different origins; in refineries and near consumption locations. [1]

The storage of oil, gas or refined products is above all a strategic necessity (Strategic Petroleum Reserve); it has been widely developed, particularly in the USA and Europe, after the OPEC embargo in 1973, following the Ramadan (Kippur) war between the Arab countries and Israel. It aims to guarantee energy independence for 2 to 3 months, or even longer, in case of oil or gas market fluctuations. The storage also allows to regulate the local market, especially during seasonal variations, and climatic changes, and take advantage of variations in the price of hydrocarbons, as is the case in this period. [2]

The International Energy Agency, which is part of the OECD, requires its members to maintain reserves equivalent to at least 90 days of net imports in the previous year. The three members that are net oil exporters (Norway, Denmark and Canada) are therefore not required to maintain reserves. In this calculation, the assessment of imports, like that of reserves, includes both crude and refined products. The OECD also specifies that this minimum volume is calculated without including oil held by end-users (so-called tertiary stocks: power plant reserves, contents of heating tanks, etc.), held in offshore tankers, reserved for the military or present in gas stations. On the other hand, reserves located outside the country under bilateral agreements may be included. Reserves in the geological sense (oil not yet extracted) are not included because they are not readily available. IEA member countries also cooperate on the use of their strategic reserves. This is for example the case in 2011 during the interruption of production due to the situation in Libya. A joint decision was taken to draw oil from these reserves. The major non-IEA countries (notably India and China) also built up strategic reserves at the beginning of the 21st century. In 2020, the total global oil storage capacity is estimated by Rystad Energy to be 4.5 billion barrels. This includes strategic reserves available to states and corresponds to 45 days of global consumption (100 million barrels per day on average in 2019). [3]

In Africa, the notion of strategic reserves is still neglected because most countries are producers and they have many sources of renewable energy without forgetting that seasonal variations in Africa are negligible due to the absence of winter. Nevertheless, we note that more and more projects for the creation of storage centers for petroleum products are emerging. In Cameroon, the SCDP remains the main actor in the storage of petroleum products but more and more private companies are investing in the creation of storage centers. The underground storage of hydrocarbons was born in Canada in 1915 and has developed everywhere in the world except Africa. Today, there are more than 600 underground natural gas storage sites with a useful volume of 310 billion m^3.

1. Main Objective

The objective of this work is to simulate the underground storage of gas in the case of depleted oil layers and to highlight the main mechanisms of flow and storage of gas in the reservoir.

2. Specific Objectives

- Simulate the production of the reservoir until it is exhausted;
- Simulate the process of injection and storage of gas in the reservoir.

3. Plan of the brief

This work is organized in 3 chapters, namely the generalities on the storage, the materials and methods which allowed to lead to the interpretation of the results of the simulation of the underground storage of gas in depleted oil beds.

Chapter I: General

The storage of energy resources is necessary to compensate for fluctuations in supply due to all kinds of hazards in production, transport and refining or variations in consumption, which depend in particular on weather conditions. It is also strategic to ensure a minimum of energy autonomy for the consumer country. Storage must be ensured at the various stages of the oil journey, from the production well to the consumption sites. This storage concerns crude oil, feedstock, intermediate cuts and finished products before shipment.

1. Storage facilities

Tanks are the most used storage means in the oil industry. They can be cylindrical or spherical depending on the product to be stored. Tanks, generally of cylindrical shape, are of three types:

- Fixed roof tanks are used for the storage of unstabilized oil (i.e. oil that still contains volatile hydrocarbons that can outgas). There are two types of fixed roofs, conical roofs and domed roofs (spherical or ellipsoidal).
- Floating roof tanks (single deck or double deck), used for the storage of stabilized oil (no risk of degassing). The roof floats on the stored product and seals with the tank shell by means of a gasket.
- Flexible tanks are another means that is just as watertight and reliable as the previous alternatives. The capacity of this type of product is very important and can go up to 1500 m^3. Its use tends to develop today in relation to the sustainable development approaches of companies. [4]

For the LPG case, we distinguish:

- cigars
- the spheres.

2. Methods of storage

For consumers, storage guarantees a continuous supply. Thus, the product circulating towards the consumption areas is not necessarily used immediately. It can then be stored to be reused as soon as the demand justifies it. In general, there are 3 types of storage, namely

2.1 Aerial storage

Aboveground storage does not require any particular geological conditions. It represents the vast majority of the large capacity tanks containing flammable liquids. When they are above ground, they are most often metallic and vertical. This type of storage is characterized by the fact that the tank is located entirely above the surrounding ground.

2.2 Covered storage

Large capacity covered tanks containing flammable liquids represent a very small portion of the tank fleet. They are mainly found in military areas or at some oil tankers. Three types of covered tanks can be identified:

- underground tanks in caves ;
- rockfill tanks ;
- buried tanks.

2.2.1 Underground tanks in caves

The principle consists in placing the reservoir in a cavern built for the purpose (figure 1). In most cases, the vault of the cavern is made of concrete and is about one meter thick. It is covered by several meters of backfill. The safety objective assigned to the "cavern" protection is in this case dictated by military safety concerns, i.e. protection against mechanical effects (from overpressure and missiles).

Figure 1: Cave tank.

Source: French Butane Propane Committee.

2.2.2 Rockfill tanks

In this case, the reservoirs are located within a basin built in excavation or in rockfill, the walls of which are bricked up against the rock. The basins are then covered with a concrete vault which can be more than one meter thick. Such an arrangement can concern a tank or a group of tanks.

2.2.3 Underground tanks

When the tanks are buried (figure 2), they have the particularity of being horizontal and toroidal in shape. They are usually made of metal, but can also be made directly of concrete. When they are made of concrete, the thickness of the wall can be greater than 1 meter. In this case, military security requirements dictate the thickness of the wall.

Figure 2: Double-walled underground tanks.
Source: Ferrero Containers.

2.3 Storage under slope

The first storage tanks under slope (Figure 3) were built for the most part in Germany, a country that opted for this technique at the beginning of the 1980s, notably under the impetus of Professor MANG of the University of Karlsruhe. This type of storage is an intermediary between aboveground and covered storage.

Figure 3: Under slope tanks.
Source : Bocom Gaz.

2.4 Underground storage of hydrocarbons

2.4.1 Storage in mined tunnels

By conventional means, galleries similar to mining galleries or tunnels are dug, but these galleries are closed, except for the tubes which allow the circulation of the stored fluids. The excavation cost is relatively high, which makes this technique attractive only for liquid or liquefiable products.

Figure 4: Digging of an unlined storage gallery at Lavéra.

Source: Geostock photo library

2.4.2 Salt cavern storage

In this case, the remarkable solubility of rock salt is used to dig large cavities in deep salt deposits by simply circulating fresh water. This technique can be used for all liquid, gaseous or liquefied hydrocarbons

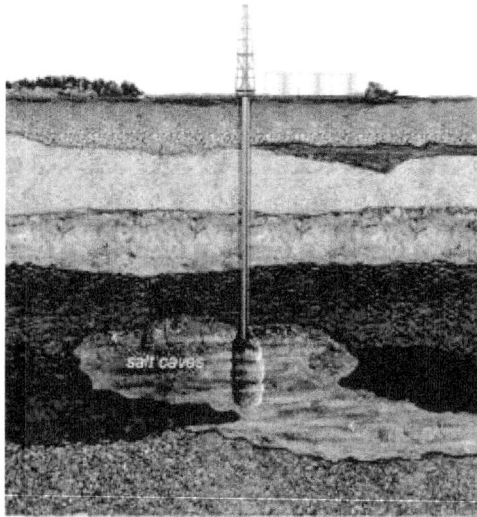

Figure 5: Saline cavity.

Source : Gazprom

2.4.3 Storage in aquifers and depleted fields

We reproduce a natural gas deposit by choosing a site that meets all the conditions, i.e. a sufficiently porous and permeable level, topped by a thick impermeable layer of curved shape, so that the gas is trapped as if under a bell in the upper part of the structure.

Instead of being abandoned at low pressure when the hydrocarbons can no longer be exploited, the gas or oil deposit is relined to allow for cycling (injection and withdrawal) of the gas. When it is an oil field, the gas withdrawn has been enriched in condensates (by vaporization under pressure at the bottom of the lightest compounds of the oil), which can be an additional interest for the storage exploitation. [5]

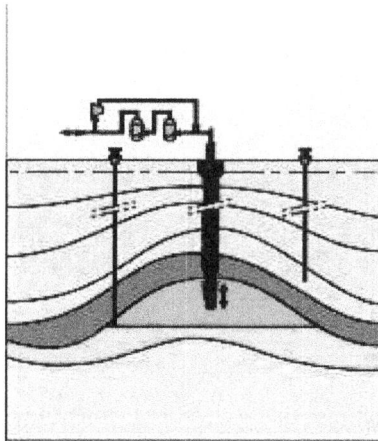

Figure 6: Aquifers & depleted fields.
Source : ©Phototheque Géostock

Figure 7: Depleted fields.

Source: Gazprom.

Table 1: Characteristics of the different types of underground storage.

Features	Mined cavity	Saline cavity	Storage in aquifers and porous media
Waterproofing	hydrodynamics	intrinsic	intrinsic (cover and bottom)
Setting up	Excavated (machines / explosives)	Leached (fresh water)	Natural voids (porous network)
Stored products	Natural gas, crude and refined oil and compressed air	Liquid hydrocarbons (crude oil, naphtha, LPG), compressed air and hydrogen	Natural gas (high pressure: 30 bar to 200 bar) and CO_2
depth	Shallow	Variable depth (300 to 2000 m)	Depth : 500 to 3000 m

8

Form	Horizontal development	Vertical development	Closed structures in antiform / bell
Dimension	Height 15 to 30m, width 10 to 25 m, length 100 to 1000 m, section 80 to 650 m².	Height 100 to 300 m, diameter 30 to 70 m	Natural network of kilometer extension (similar to natural hydrocarbon reservoirs)
Volume	Volume 100 000 to 500 000 ^{m3} (10 000 - 1 000 000 ^{m3})	Volume 150 000 to 650 000 m3	Useful volume 0.5 to 4 billion Nm3 plus a gas cushion
Access	Physical access by tunnel or shaft	Only access to drilling (leaching & exploitation)	Several drillings (exploration, exploitation & follow-up)
Recognition	drilling & direct observations during excavation	drilling and indirect methods (3D image by sonar leaching)	drilling & indirect methods, geophysics, logs, tests
Operation	bottom pumps, water curtain monitoring, piezometric monitoring	no bottom equipment, creep monitoring, traps, brine management	no background equipment, interface tracking, peripheral control

Chapter II: Materials and methods

Historically, depleted gas reservoirs have been the most important and commonly used formations for natural gas storage. A depleted field is generally the most appropriate option because it has proven characteristics for storing gas. There is a considerable level of data already available on the geological, structural, and petrophysical characteristics of the primary production. Commercially, depleted reservoirs can provide very good storage efficiency, volume and peak performance in terms of injection rates and/or gas withdrawal.

1. Materials

Since this work is a simulation, the essential tool is a laptop in which we used Matlab 2019 and CMG (Computer Modelling Group) software. CMG simulates simple to more advanced recovery processes through a combination of easy-to-use model creation workflows, state-of-the-art performance enhancement technology, and interdisciplinary multi-physics effects

9

(thermal effects, geochemistry, geomechanics, fluid and phase behavior, wellbore hydraulics, and completions) needed to accurately model recovery processes.

CMG is segmented into several sub-programs with specific functionalities which are :

- CMOST: a powerful sensitivity analysis, history matching, optimization and uncertainty tool to maximize recovery from all types of reservoirs and recovery processes;
- IMEX: Model primary and secondary recovery techniques for conventional and unconventional oil/gas reservoirs; use fast and simple workflows to create forecasts with confidence;
- COFLOW: a multi-fidelity, multi-disciplinary and collaborative modeling environment for making informed decisions on large integrated oil and gas projects;
- RESULTS: Advanced visualization and analysis capabilities provide insight into reservoir characteristics, recovery processes and reservoir performance;
- GEM: the world's leading reservoir simulation software for compositional and non-conventional reservoir modeling;
- STARS: Accurately model the physics of all in-situ recovery processes - thermal, chemical, or other advanced emergency recovery techniques to maximize the value and production of an asset;
- BUILDER: The interactive, intuitive and easy-to-use interface allows quick and efficient design and preparation of simulation models for all CMG simulators;
- WINPROP: Create optimized fluid property descriptions for CMG simulators and black oil fluid property data for third-party reservoir simulation software.

In this work, we used the BUILDER and IMEX interfaces. [6]

MATLAB combines a desktop environment designed for iterative analysis and design processes with a programming language for expressing mathematical equations directly as arrays and matrices.

2. METHODOLOGY

To realize this model, the following parameters were simplified:

- the action of the temperature in the tank is neglected;
- the bubble pressure is constant;
- the compressibility of the rock is constant;

10

- the injection and production parameters are set at the limits;
- the period intervals of the operations are annual;
- the injected water is non-saline;
- The gas injected is composed of 12% ethane, 38% propane and 50% butane.

This simulation takes place in 3 steps:

- the design of the model
- primary and secondary recovery
- gas injection, storage and production.

2.1 Presentation of the model

This is a reservoir with a normal fault composed of 3 layers of constant porosity at 30%.

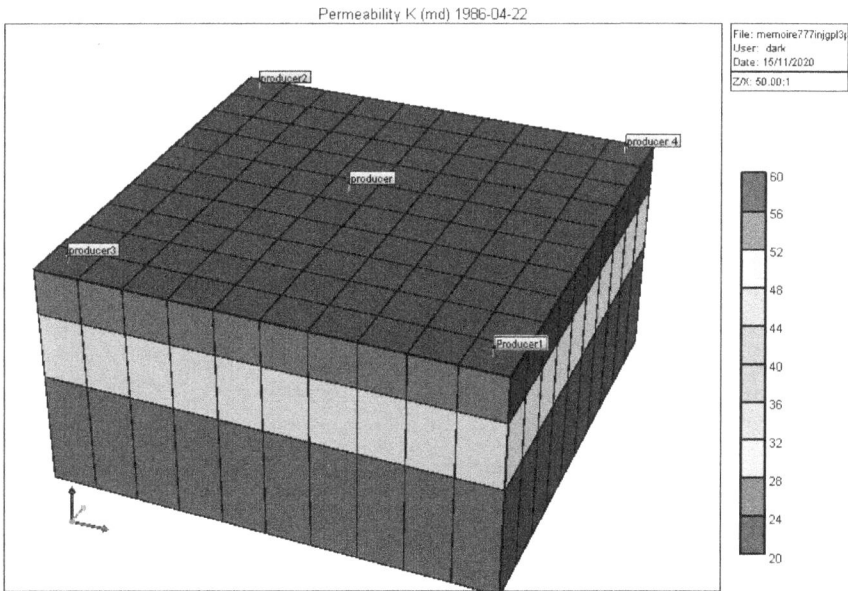

Figure 8: Distribution of permeability along the vertical.

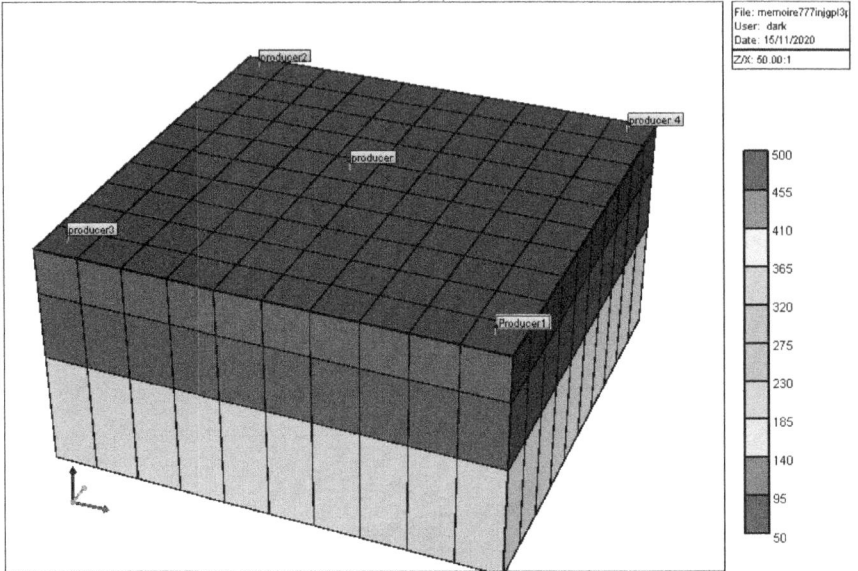

Figure 9: Distribution of permeability according to layers.

The fluid flow model here is of the "Black Oil" type, i.e. the fluid flow equations are partial differential equations in contrast to the compositional formulation where the hydrocarbon components are treated separately. This term refers to a fluid model, in which water is explicitly modeled with two hydrocarbon components, an oil phase and a gas phase.

The equations for an extended black oil model are: [7]

$$\frac{\partial}{\partial t}\left[\varphi\left(\frac{S_O}{B_O}+\frac{R_V S_g}{B_g}\right)\right] + \nabla.\left(\frac{1}{B_O}\overrightarrow{u_g}+\frac{R_V}{B_g}\overrightarrow{u_g}\right) = 0$$

$$\frac{\partial}{\partial t}\left[\varphi\left(\frac{S_w}{B_w}\right)\right] + \nabla.\left(\frac{1}{B_w}\overrightarrow{u_w}\right) = 0$$

$$\frac{\partial}{\partial t}\left[\varphi\left(\frac{R_S S_O}{B_O}+\frac{S_g}{B_g}\right)\right] + \nabla.\left(\frac{R_S}{B_O}\overrightarrow{u_o}+\frac{1}{B_g}\overrightarrow{u_g}\right) = 0$$

Where ϕ is the porosity, Sw is the water saturation, So, Sg are the liquid and vapor phase saturations in the reservoir. Oil and gas at the surface (standard conditions) could be produced from both liquid and vapor phases. This is characterized by the following quantities:

- Bo is a factor of the formation oil volume (ratio of a certain volume of tank liquid to the volume of oil under standard conditions obtained from the same volume of tank liquid),
- Bw is a factor of the formation water volume (ratio of water volume at tank conditions to water volume at standard conditions),
- Bg is a factor of the formation gas volume (ratio of a certain volume of tank vapor to the volume of gas at standard conditions obtained from the same volume of tank vapor),
- Rs is the ratio of a gas solution to oil phase (ratio of gas volume to oil volume at standard conditions obtained from a certain amount of liquid phase at reservoir conditions),
- Rv is a ratio of vaporized oil to gas phase (ratio of oil volume to gas volume at standard conditions obtained from a certain amount of vapor phase at tank conditions).

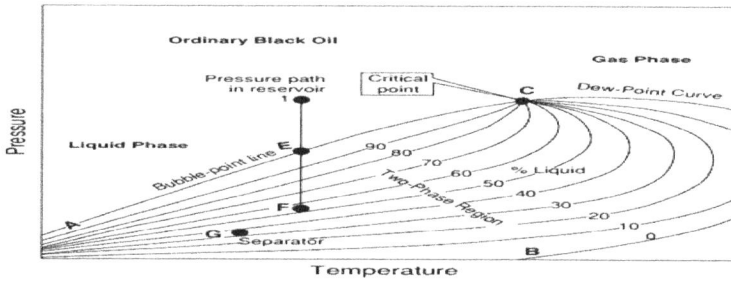

Figure 10: Typical phase diagram of black oil.

Source: "PVT Properties of Black Crude Oil

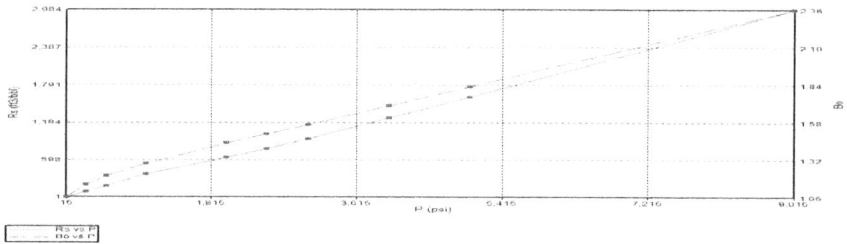

Figure 11: Relationship between solution gas ratio, formation oil volume factor and pressure.

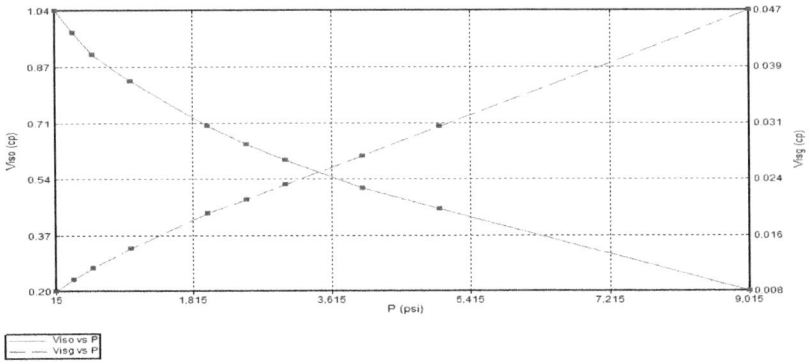

Figure 12: Relationship between oil and gas viscosity as a function of pressure.

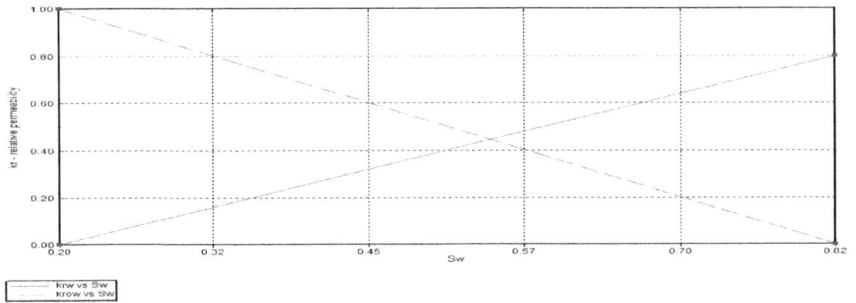

Figure 13: Relationship between relative oil and water permeability as a function of water saturation.

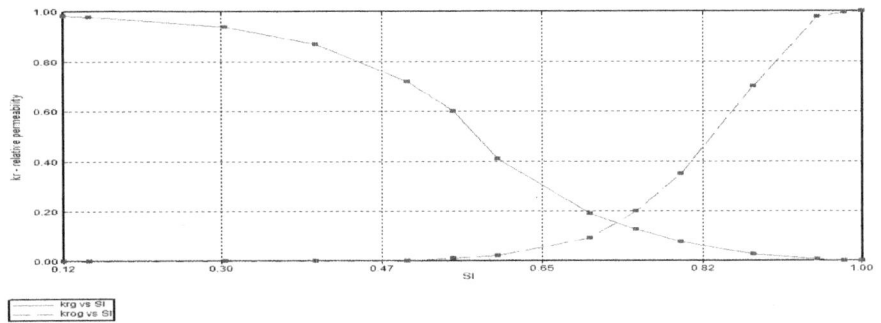

Figure 14: Relationship between relative permeability of oil and gas as a function of water saturation.

14

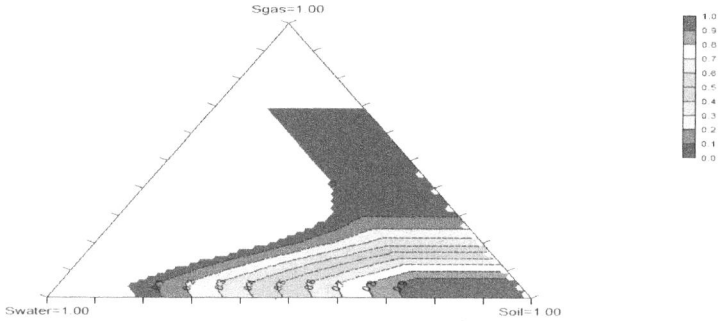

Figure 15: Relationship between water, oil and gas saturation.

2.2 Primary and secondary   recovery

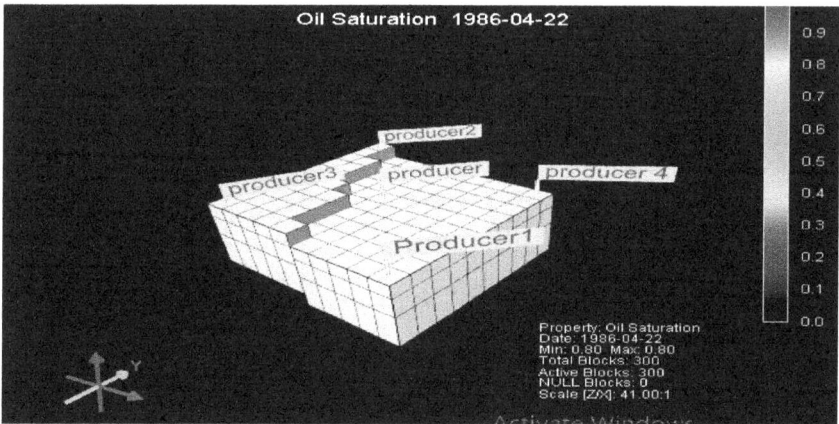

Figure 16: Oil saturation initially in the tank.

Black oil reservoirs that produce exclusively by a solution gas mechanism usually recover 10 to 25% of the OOIP by pressure depletion, but in our case it is a combination of several drainage mechanisms.

The production start time is set for April 22, 1986. 5 wells were drilled as shown in Figure 15 to optimize the primary recovery and then the producer wells 3 and 4 will be replaced by water injectors as secondary recovery to ensure that the reservoir is depleted.

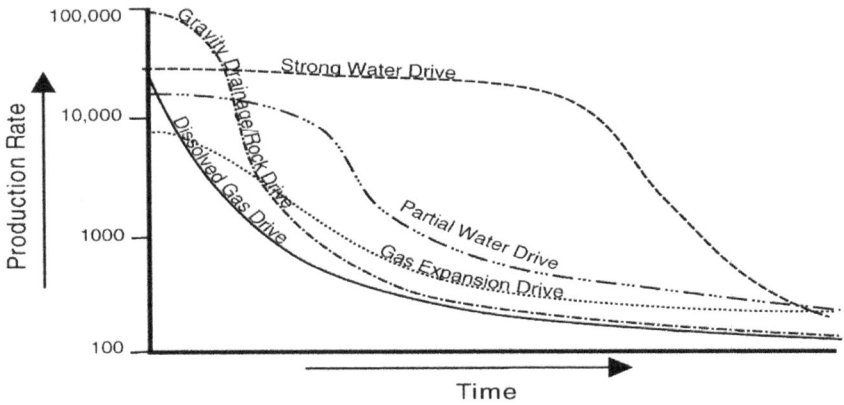

Figure 17: Declination curves according to the drive mechanism.

Table 2: Tank parameters.

Density of the oil	46.244 lbm/ft3
Density of the gas	0.0647 lbm/ft3
Density of water	62.238 lbm/ft3
Water-oil contact depth	9500 ft
Gas-oil contact depth	7000 ft
Compressibility of the rock	3.0E-6
Bubble point pressure	4014.7 Psi
Tank pressure	4805 KPa
Volume of oil initially in place	0.25858E+09 STB
Volume of gas initially in place	0.32839E+12 SCF
Volume of water in place	0.10564E+09 STB

Hydrocarbon pore volume	433591 M RBBL
Total pore volume	541989 M RBBL

- **Machine Learning modeling of gas phase expansion volume determination during primary and secondary recovery**

The simulation model is designed on the following assumptions:

- the influence of temperature is negligible;
- the influence of the aquifer is negligible;
- the value of the bubble pressure is constant;
- the fluid is initially monophasic in the tank.

Therefore, the goal here is to determine through a computer model, the variation of gas saturation as a function of the evolution of water saturation, oil saturation and reservoir pressure during primary and secondary recovery. The algorithm used is the neural network (Bowie, 2018) via the "Neurol Net Fitting" module on Matlab 2019.

As a result, the data was divided into three sets:

- the training data used to model the magnitude up to 70%;
- test data to check the robustness of the model at 20%;
- validation data to confirm authenticity equivalent to 10%;

The following graphs show the different scores obtained during the validation of the model.

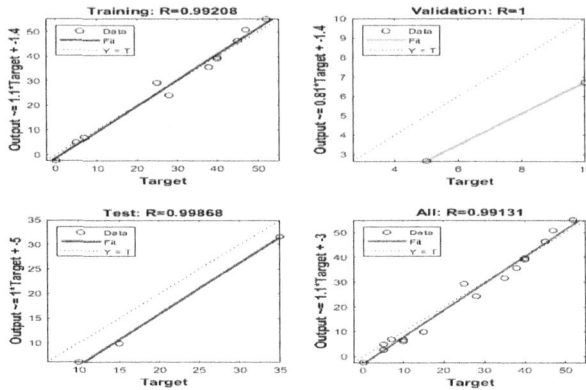

Figure 18: Validation of the computer model

The different curves show:

- for training data: the score obtained is 0.992
- the tests are 99.8 % conclusive
- the validation provided 100% satisfactory results

These different results show that the model obtained (Appendix 12) is satisfactory, so it can be used for the prediction of gas saturation for this type of model.

2.3 Gas injection, storage and production

After the secondary recovery, depending on the drainage mechanisms, a recovery of about 40 to 70% is possible so that we can start injecting gas. The composition of the injected gas is 12% ethane, 38% propane and 50% butane. It is LPG (liquefied petroleum gas). The base gas, or cushion, can include both native and injected gas. This is the volume that must be left in the underground storage to provide the required pressurization so that the remaining gas can be delivered at reasonable rates while the amount of gas injected and withdrawn during the normal storage cycle is called working gas or head gas (70%).To inject gas into the reservoir, the previous water injector wells will be converted back to gas injectors as well as the wells in the center and bottom wing of the model. Gas will be injected and stored in the reservoir so that we can observe the conservation mechanism and the behavior of the gas during production once the head gas is in sufficient quantity.

The following approximations were made for volume estimation: 1bbl =5.61 SCF, 1STB= 0.16 m^3, 1SCF= 0.026 m^3, 1 m^3 = 6.28 bbl, 1 bbl=0.158 m^3. [10]

Chapter III: Results and discussions

1. Primary recovery

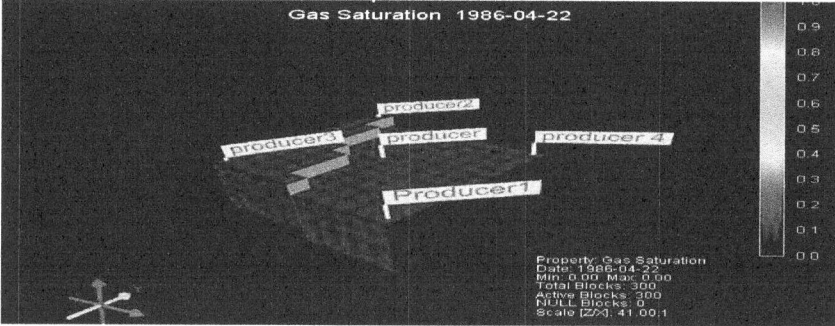

Figure 19: Initial distribution of gas saturation.

We note that initially in the tank, there is little or no free gas due to the conditions of the tank as represented in Figure 10 precisely because of the high bubble pressure in the tank.

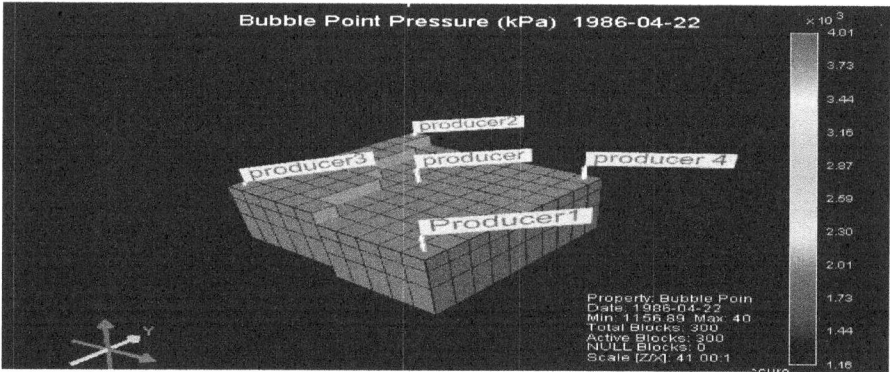

Figure 20: Initial distribution of bubble pressure.

During the natural depletion of the reservoir, a cumulative production of oil of 28871 MSTB, gas 264779MMSCF and water 35,336MSTB is recorded, which makes 229684 MSTB of oil, 63473 MMSCF of gas and 105603 MSTB of water remaining in the reservoir. The percentage of hydrocarbon recovery is on average 46% (12% oil and 80% gas) due to the entrainment of gas in solution in the oil and the aquifer. The simulation duration is 11 years. The bubble pressure decreases significantly in the reservoir during production and a significant increase in gas saturation in the reservoir is observed.

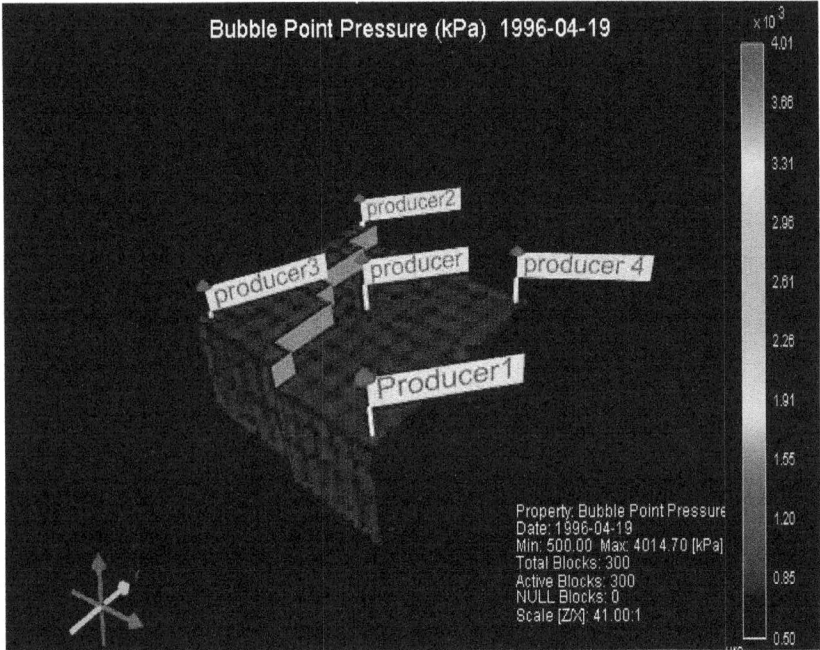

Figure 21: Distribution at the end of the primary recovery of the bubble pressure.

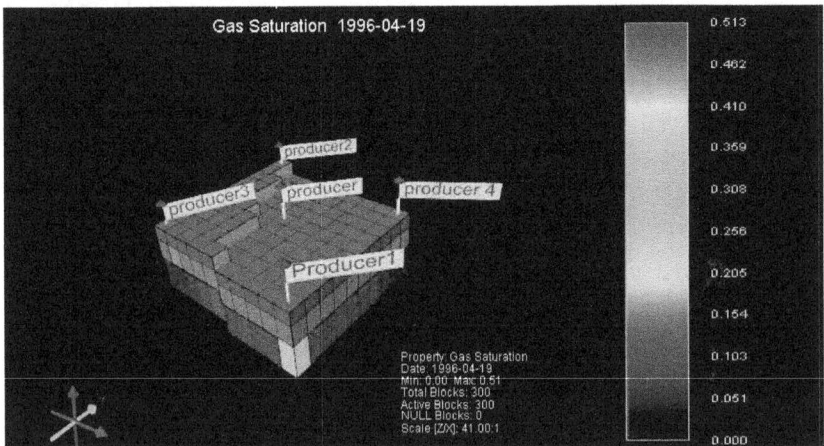

Figure 22: Distribution at the end of the primary recovery of gas saturation.

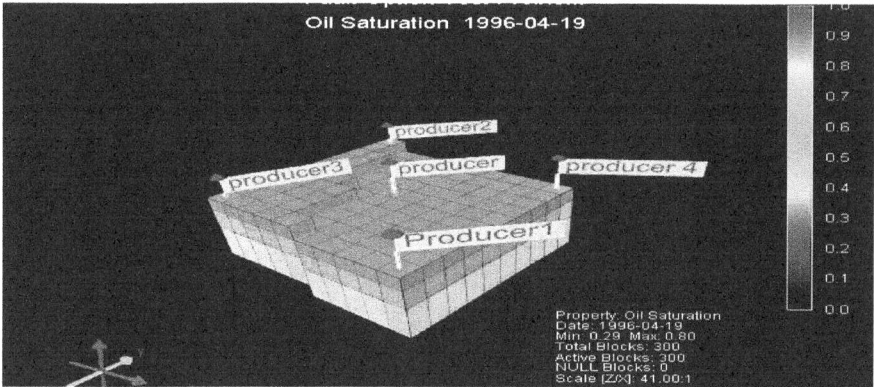

Figure 23: Distribution at the end of primary recovery of oil saturation.

2. Secondary recovery

After the primary recovery, the deposit still has 54% of hydrocarbons. In order to deplete it, we inject water through the 'producer3' and 'producer'4' wells and close the 'producer2' and 'producer1' wells. It would have been possible to inject through the 'producer2' and 'producer1' wells, but the barrier fault prevents the diffusion of water through well 2, so to ensure a symmetrical sweep of the oil towards 'producer' we also close well 1.

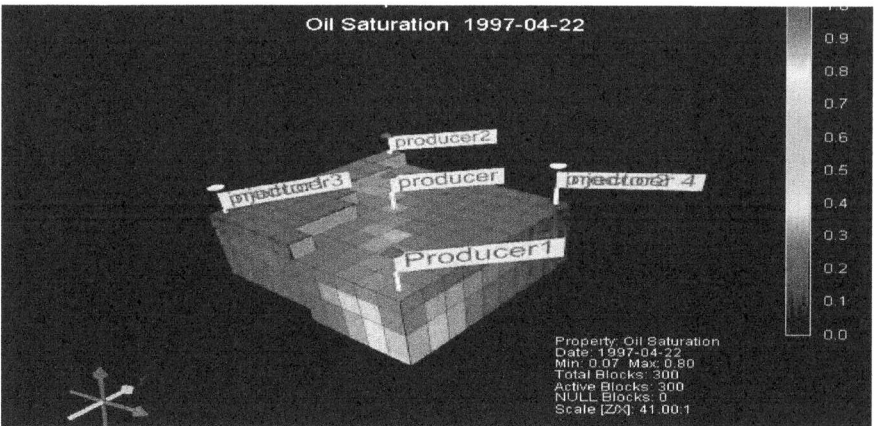

Figure 24: Oil sweeping by water.

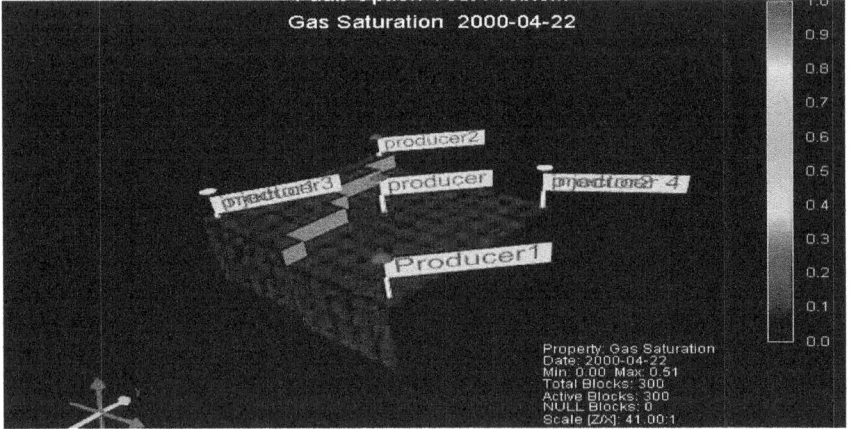

Figure 25: Gas saturation distribution at the end of the water injection.

The injection of water into the reservoir leads to a rapid decrease in gas saturation in the reservoir. The simulation time of the secondary recovery is 5 years and 3090 MMSTB of water was injected for an average recovery of 64% hydrocarbons, 18% more than the primary recovery (41% oil and 81% gas).

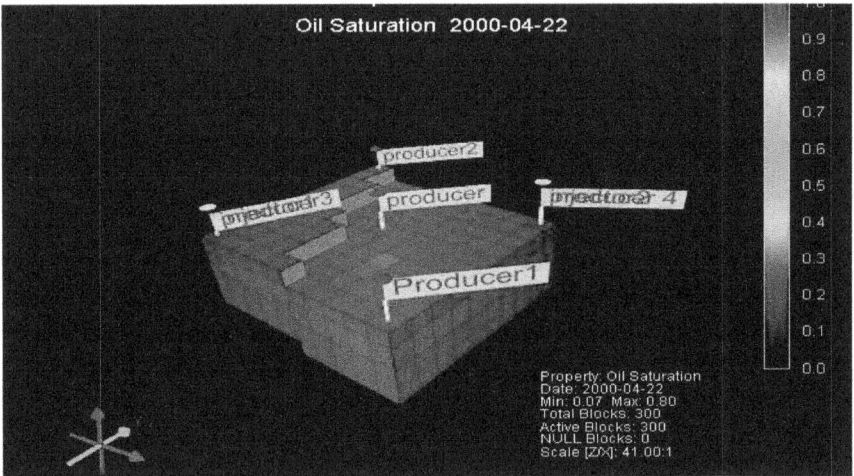

Figure 26: Distribution of oil saturation at the end of water injection.

3. Gas injection, storage and production

The reservoir being depleted, we replace the 4 still active wells by gas injectors. Neglecting the chemical reactions in the reservoir as well as the influence of temperature, we can determine the average storage capacity of the gas in the reservoir thanks to the principle of conservation of mass which is summarized as follows:

Storage capacity = Quantity of hydrocarbons produced +Quantity of water produced - Quantity of water injected

The reservoir having produced 64% of its hydrocarbon reserve, the volume extracted is estimated at 108.6 MMSTB (17376000m3) of oil and 286288 MMSCF (7443488000m3) of gas. The quantity of water injected during secondary recovery is 3092 MMSTB (494720000m3) and that produced is 2708 MMSTB (433280000 $^{m3)}$ then the gas storage capacity of the reservoir is :

Storage capacity = 17376000+7443488000+433280000-494720000= 7399424000 ^{m3}

The reservoir can therefore hold an average of 28,4593 MMSCF at normal pressure conditions, but it is possible to inject more, taking into account the compressibility of the gas and the elasticity of the reservoir. This will increase the deliverability of the producing wells but could damage the stability of the reservoir and create cracks in the caprock.

Table 3: Schedule of injection, rest and production operations.

periods	operations
2000-2010	Gas injection (4 wells)
2010-2015	Rest and observation of the tank
2015-2020	Production from the central well
2020-2030	Periodic production of the central well (50%)

2030-2035	Injection from Gas wells 1,3 and 4 and production from central well
2035-2036	Production from the central well
2036-2040	Production and periodic injection from the central well
2040-2042	Central production
2042-2045	Rest
2045-2050	Gas injection by the 4 wells
2050-2053	Production by the 4 wells

The Producer 2 well is closed after the primary recovery because its position near the fault limits its connection with the other blocks of the model, which leads to errors during the simulation.

a) 2000-2010 : Gas injection (4 wells)

After water injection, an increase in reservoir pressure is observed as shown in the following figures.

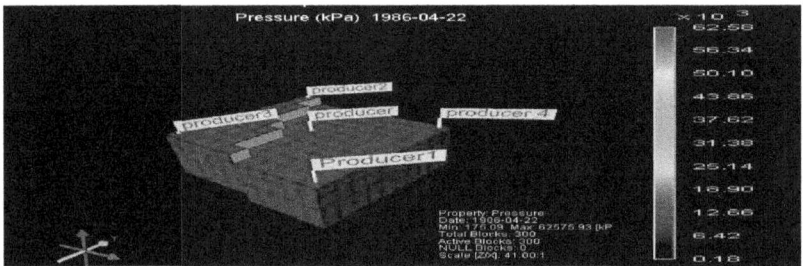

Figure 27: Distribution of initial pressures in the reservoir.

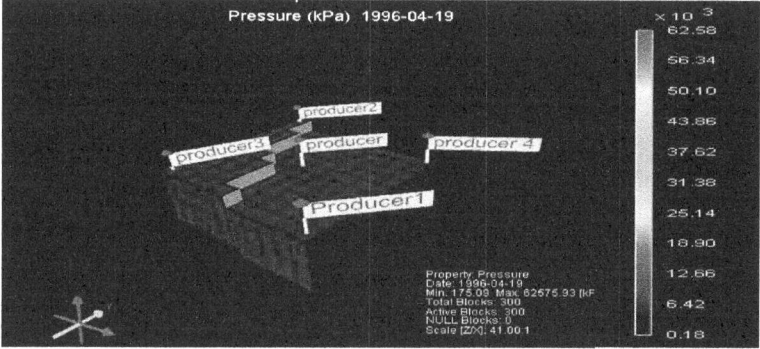

Figure 28: Pressure distribution after primary recovery in the tank.

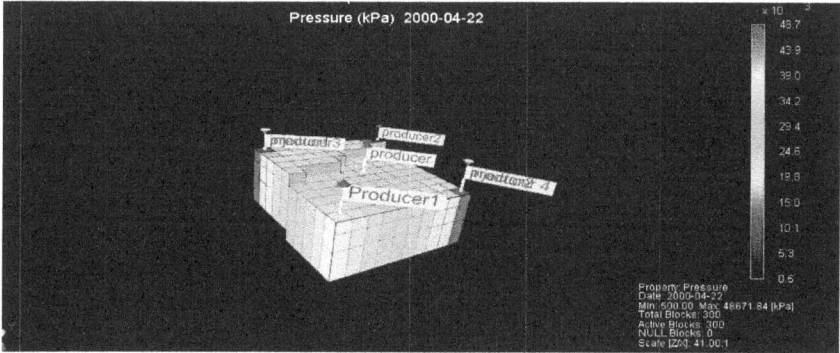

Figure 29: Pressure distribution after secondary recovery in the tank.

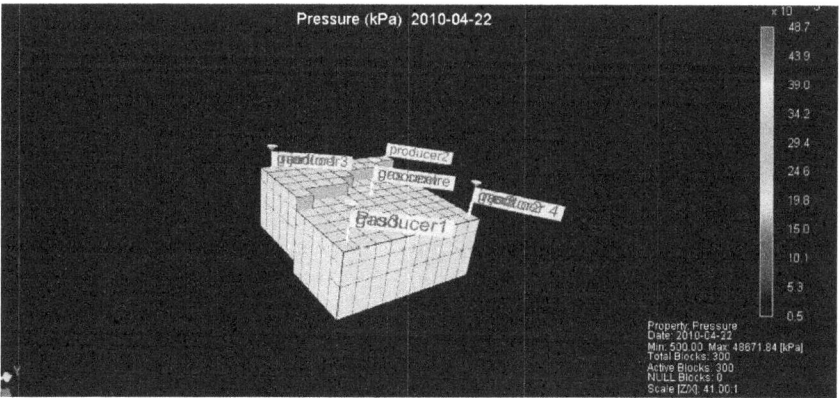

Figure 30: Pressure distribution after gas injection in the reservoir.

25

The difference between the pressure distribution after water and gas injection is that the water injection was done with two diagonally opposed wells from which we observe an increasing evolution of the pressure of the injectors towards the center of the model by taking into account the mobility of water while the gas injection was done by 4 wells which results in a homogeneous distribution of the pressure.

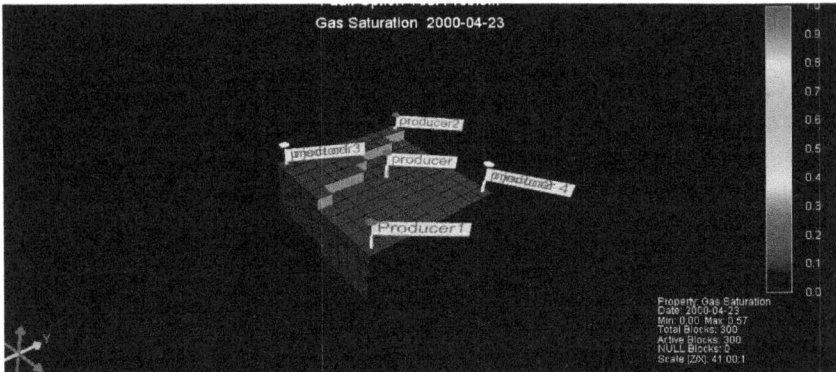

Figure 31: Gas saturation before gas injection.

The gas saturation before and after injection does not vary because the injected gas will dissolve in the residual oil. As gas is injected, there are gradual variations in gas saturation in blocks in the layers. These variations indicate the new zones of bubble pressure increase caused by the injection.

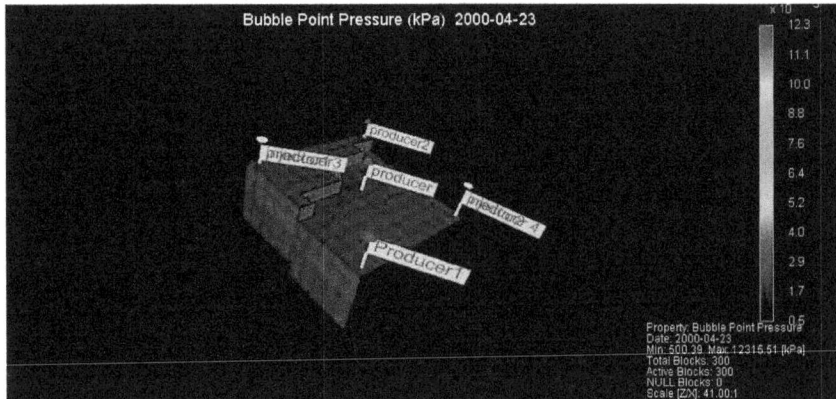

Figure 32: Bubble pressure distribution before gas injection.

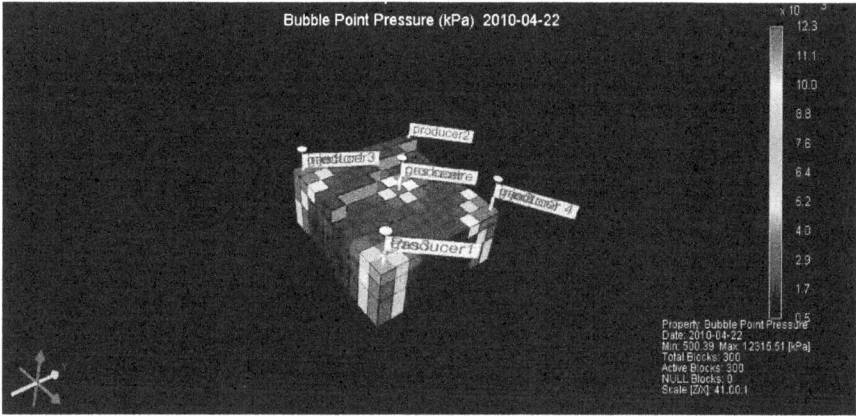

Figure 33: Bubble pressure distribution after gas injection.

This high bubble pressure in the injection zones causes the gas to become trapped in the oil, resulting in increased oil saturation in these zones. This can also result in the injected gas being in liquid form in the reservoir.

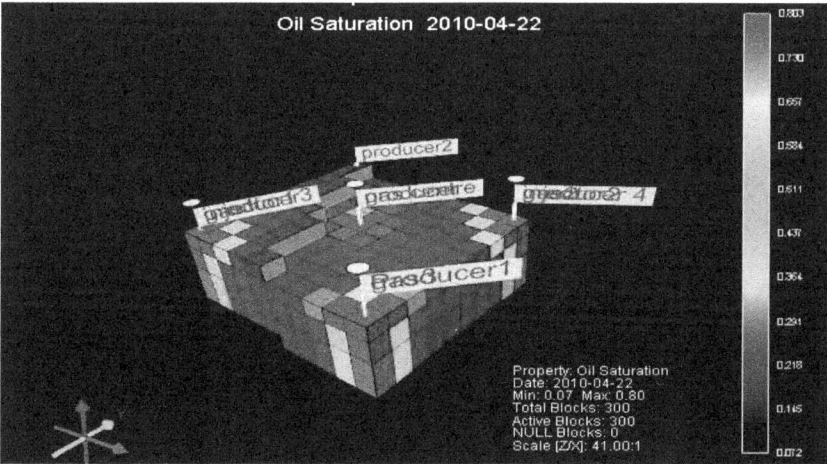

Figure 34: Distribution of oil saturation after injection.

b) 2010-2015: Rest and observation of the reservoir

During the resting phase of the model, we observe no change in pressure, bubble pressure and gas saturation, but a slow increase in oil saturation is observed, which corresponds to a dissolution of the gas injected into the oil phase due to the pressure conditions of the tank. In order to release the gas in solution, it will be necessary to create a depressurization in the tank in order to make the bubble pressure drop.

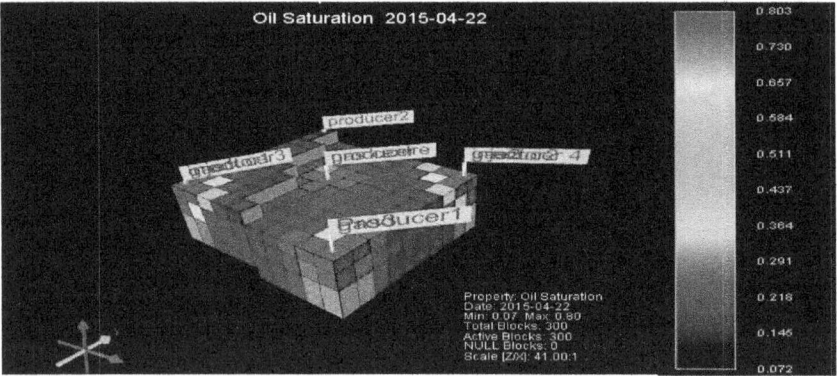

Figure 35: Distribution of oil saturation after rest.

c) **2015-2020 : Production by the central well**

In order to create a depressurization in the model, we stop the injection and open the central well. We observe a migration of the oil towards the depressurization zone and in a first time, a production of condensates and gas.

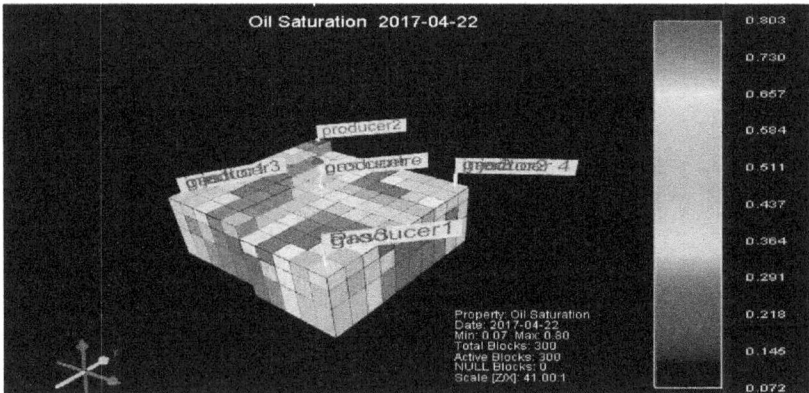

Figure 36: Distribution of oil saturation in 2017.

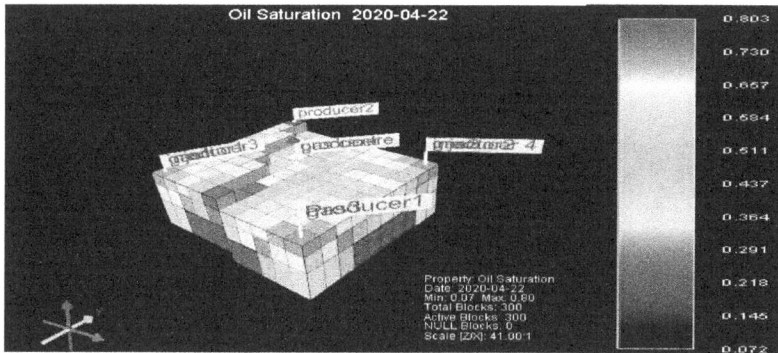

Figure 37: Distribution of oil saturation in 2020.

The production of the central well has led to a drop in bubble pressure in the oil saturated zones and we observe a strong release of gas in solution.

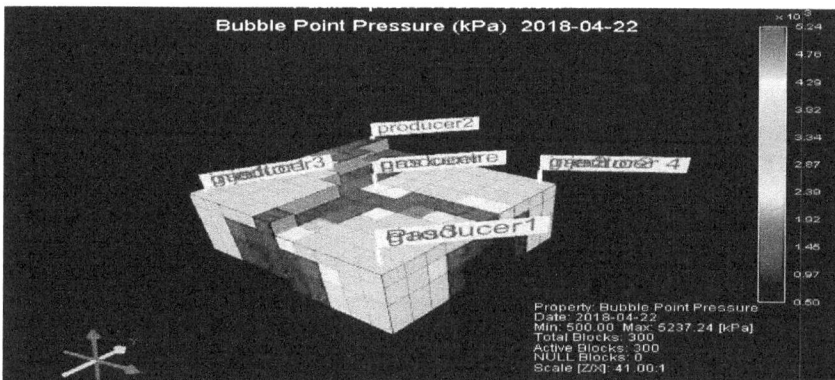

Figure 38: Distribution of bubble pressure in 2018.

Figure 39: Gas release in 2018.

A production of condensate will be observed in the central well because the oil enriched by the gas will migrate rapidly to the depressurization zone. Following this production phase, the reservoir pressure will return to normal.

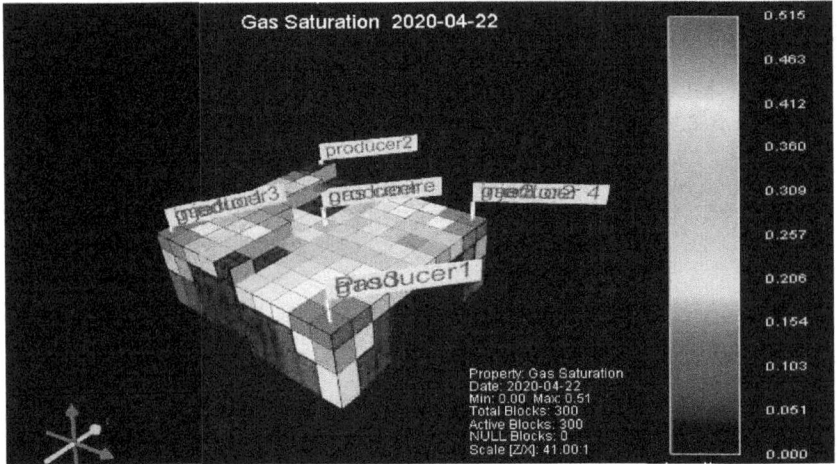

Figure 40: Distribution of gas saturation in 2020.

d) 2020-2030: Periodic production of the central well (50%)

In order to maintain the gas release, we keep the central well open at a frequency of 50% per year. During this period we observe a decrease in oil saturation.

Figure 41: Distribution of oil saturation in 2030.

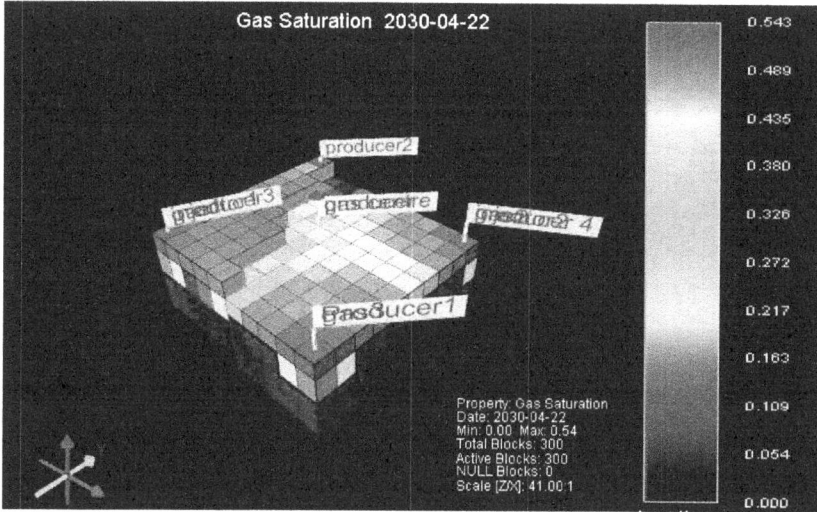

Figure 42: Distribution of gas saturation in 2030.

The change from oil to gas phase leads to a progressive increase of the gas saturation in the reservoir. It is also noted an expansion of free gas in the upper layers of the model. We have therefore moved from an initial model without free gas to a model with free gas.

e) 2030-2035: Injection from 'Gas' 1,3 and 4 wells and production from central well

In order to better understand the storage mechanisms of the model, the central well remains open while we inject gas into the other wells.

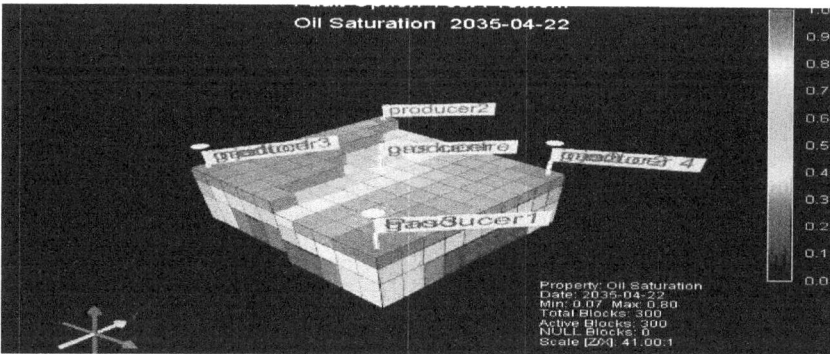

Figure 43: Distribution of oil saturation in 2035.

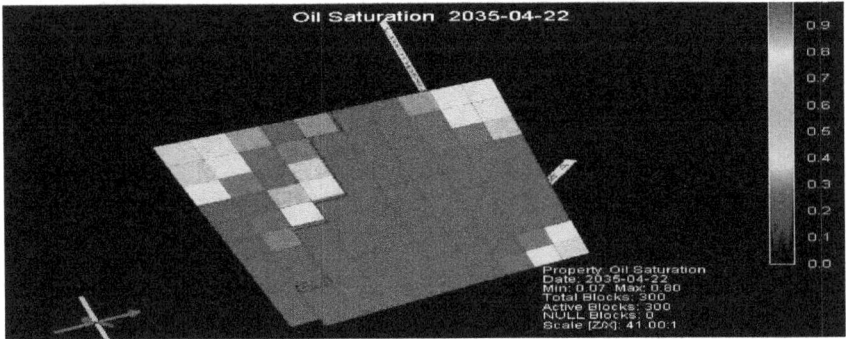

Figure 44: Distribution of oil saturation in 2035 (bottom view).

No significant evolution is observed in the upper layers of the model, but in the lower layers, the oil saturation progressively increases. This can be explained by two phenomena, part of the gas injected in the reservoir is dissolved in the oil in the intermediate and deep layers while the other part supplies the gas free layer. The gas free layer, by gaining volume, pushes the oil towards the lower layers of the model. The bubble pressure increases slightly and evenly in the reservoir while the reservoir pressure remains almost static.

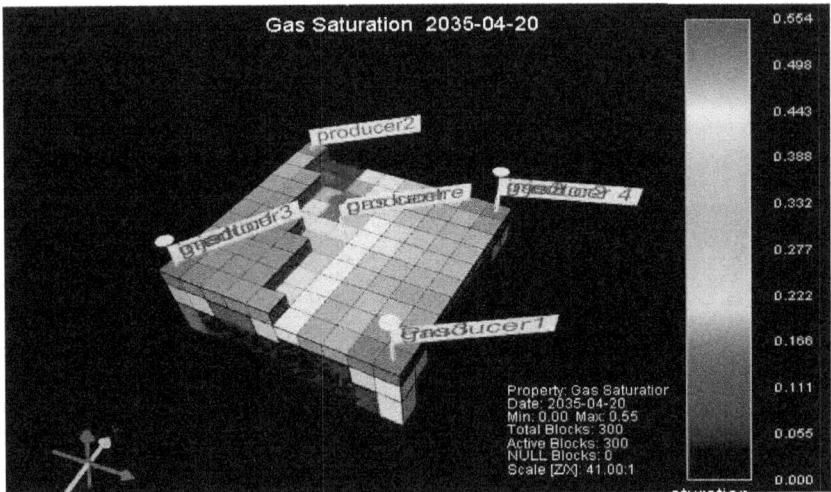

Figure 45: Distribution of gas saturation in 2035.

f) 2035-2036: Production by the central well

During this period, we stop injecting gas in order to produce from the central well.

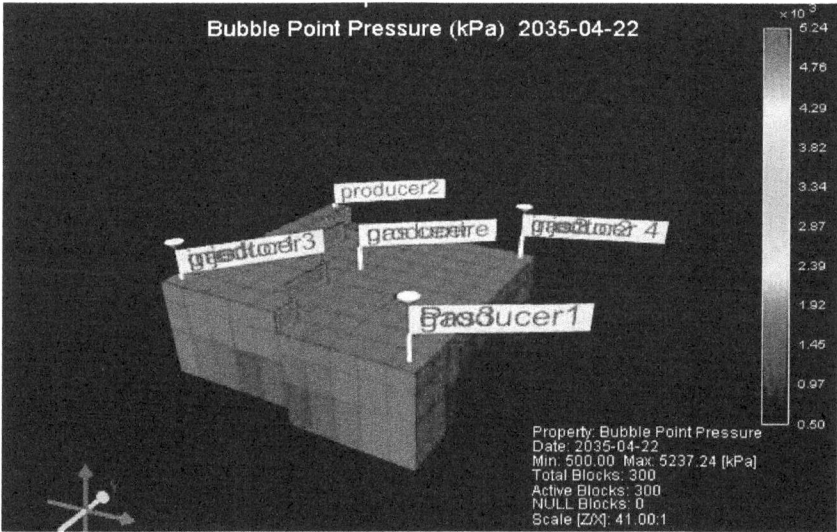

Figure 46: Distribution of bubble pressure in 2035.

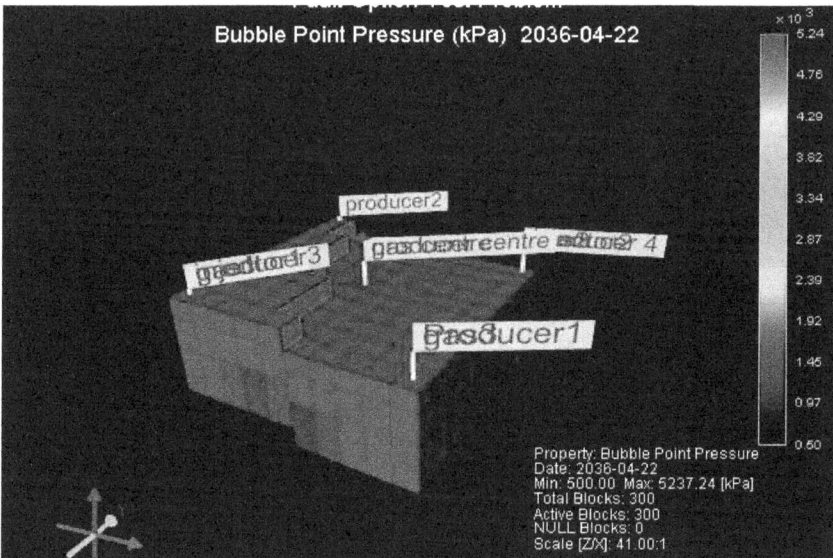

Figure 47: Distribution of bubble pressure in 2036.

The bubble pressure that had increased during the previous period falls back, which leads to a release of gas and improves the gas saturation.

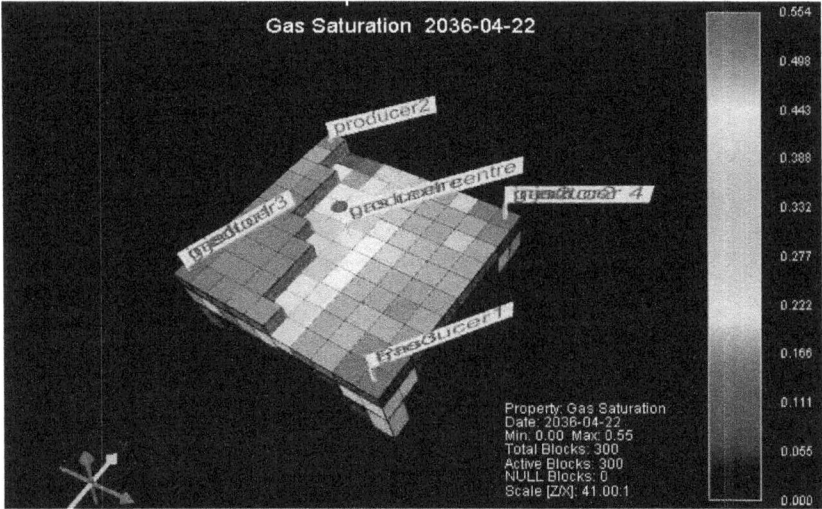

Figure 48: Distribution of gas saturation in 2036.

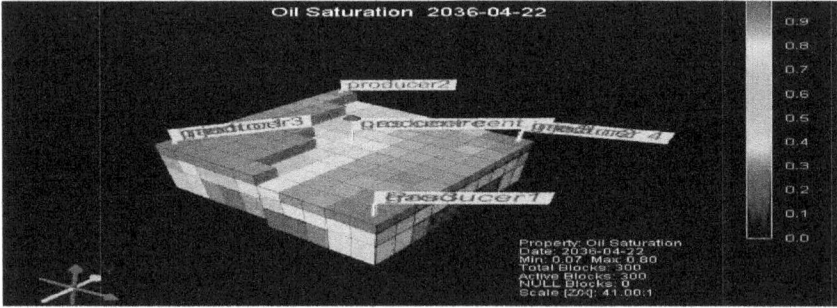

Figure 49: Distribution of oil saturation in 2036.

The oil saturation evolving inversely with the gas saturation, we note a decrease but it is to note at the level of the producing well a balance between the oil and gas saturation.

g) 2036-2040: Production and periodic injection from the central well

In 2040, following a periodic injection and production from the central well of about 50% each, an increase in gas saturation and a considerable expansion of the gas at the surface is observed.

On the other hand, oil continues to accumulate in the lower layers, particularly those where the central well is located, which validates the hypothesis related to the dissolution phenomenon.

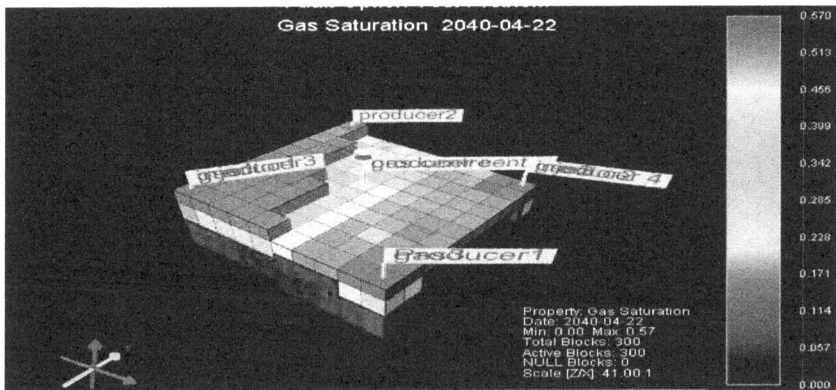

Figure 50: Distribution of gas saturation in 2040.

Figure 51: Distribution of oil saturation in 2040 (bottom view).

h) 2040-2042: Central production

The pressure drop caused by production from the central well causes a constant increase in gas saturation. This gas comes from lower areas of the model where the oil saturation is constantly decreasing.

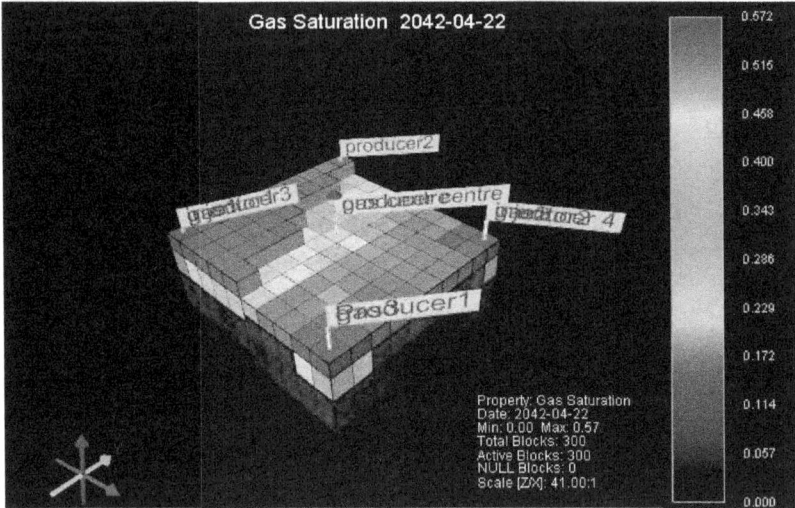

Figure 52: Distribution of gas saturation in 2042.

i) 2042-2045: Rest

The aim of this second resting phase is to observe if the gas expansion will be slowed down. The observation is the opposite of our expectations. Concentrations of oil migrate from the lower layers to the surface layers where it changes phase.

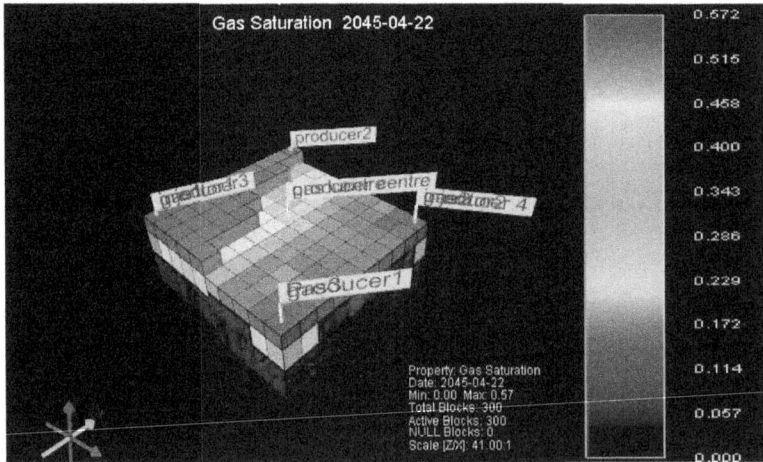

Figure 53: Distribution of gas saturation in 2045.

j) 2045-2050: Gas injection by the 4 wells

In order to verify our interpretations, we proceed to a second injection phase by 4 wells to observe if the injected gas will still dissolve in the oil and how the free gas will react.

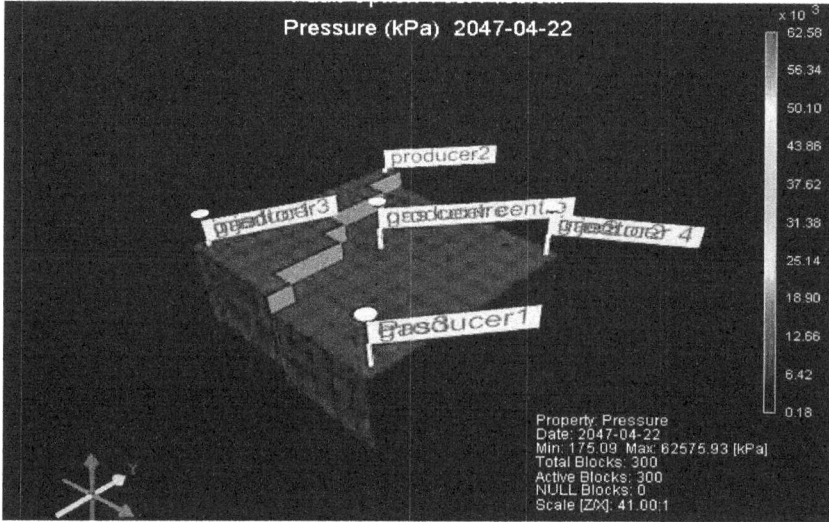

Figure 54: Pressure distribution in 2047.

The injection of gas leads to an increase in pressure in the reservoir, which facilitates the dissolution of the free gas in the oil.

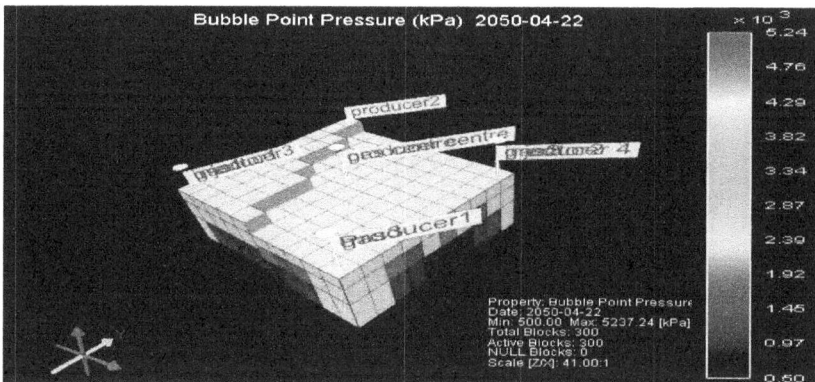

Figure 55: Distribution of bubble pressure in 2050.

The increase of the bubble pressure in the reservoir is more important at the level of the surface layers which will limit the expansion of the free gas and lead to a decrease of the gas saturation. Contrary to the first injection phase, the evolution of the bubble pressure is no longer localized at the level of the injector wells, which shows that the reservoir conditions and the nature of the fluid have evolved.

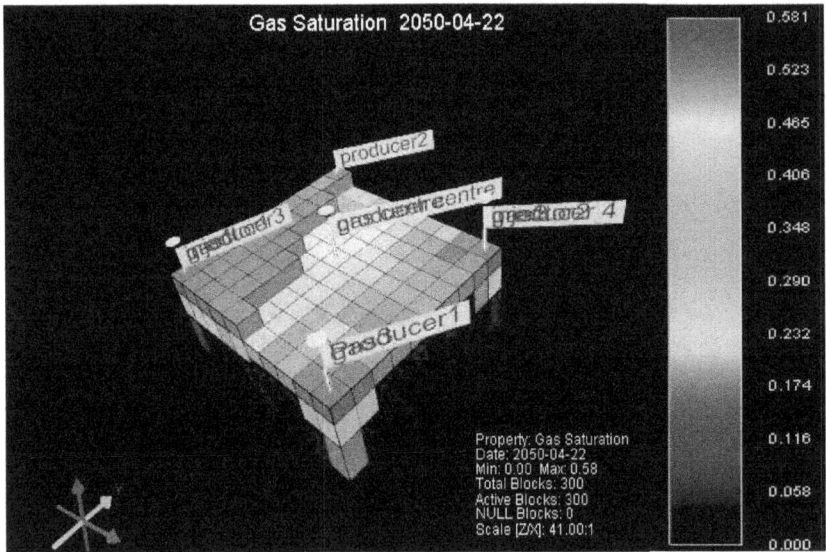

Figure 56: Distribution of gas saturation in 2050.

This decrease in gas saturation at the surface results in an increase in oil saturation in the intermediate and lower layers. The injection will therefore have the effect of increasing the pressure of the reservoir, thus creating conditions that allow the gas to dissolve in the oil or to condense, but the gas has reached significant proportions, so it is impossible to recover the state of Figure 34.

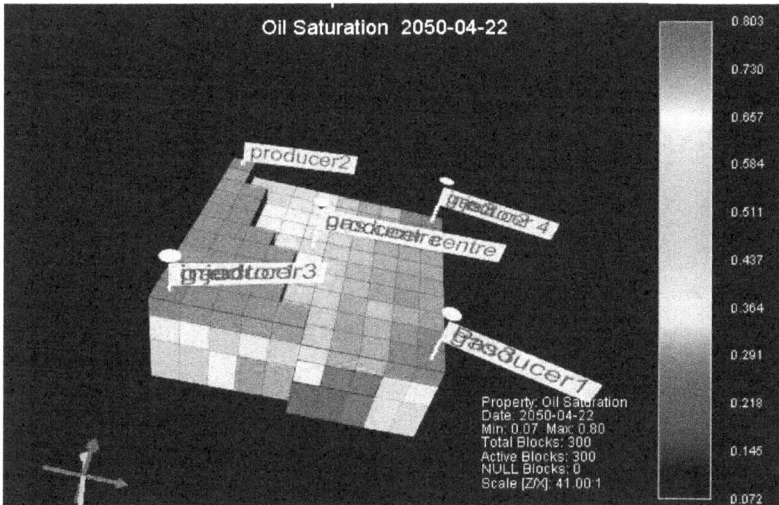

Figure 57: Distribution of oil saturation in 2050.

k) 2050-2053: Production by the 4 wells

As a final production scenario, we produced with all active wells in order to observe the consequence of a sudden depressurization of the field.

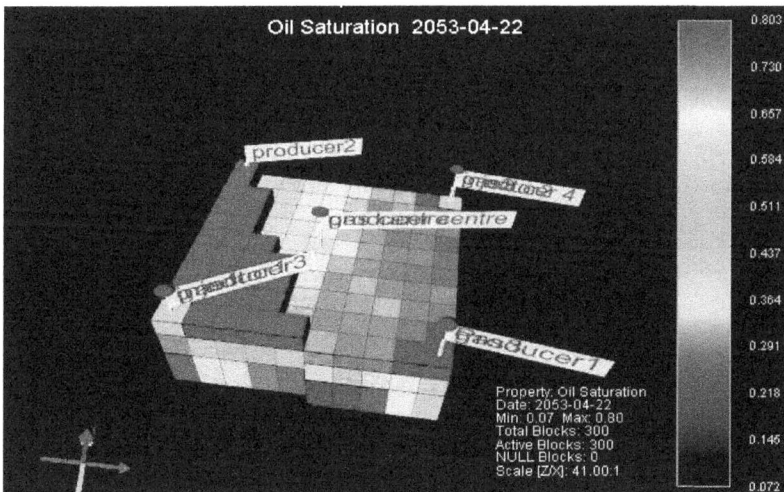

Figure 58: Distribution of oil saturation in 2053.

The start of production has led to the production of oil reserves previously stored in the layers near the injectors. A high concentration of oil has accumulated in the part of the fault close to the 'producer2' well because the suction of the central well is reduced there.

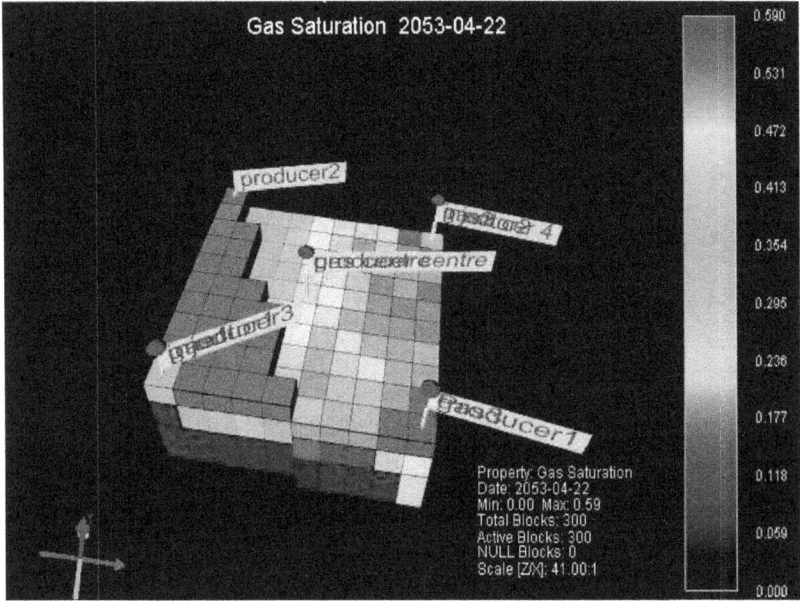

Figure 59: Distribution of gas saturation in 2053.

The opening of the production wells has led to a sudden increase in gas saturation, which is more important at the level of the previously injected wells. The bubble pressure drops considerably as well as the reservoir pressure.

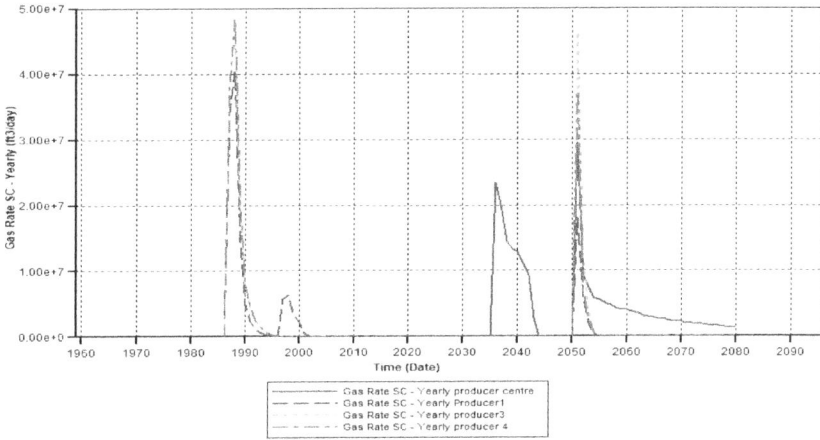

Figure 60: Gas production from wells.

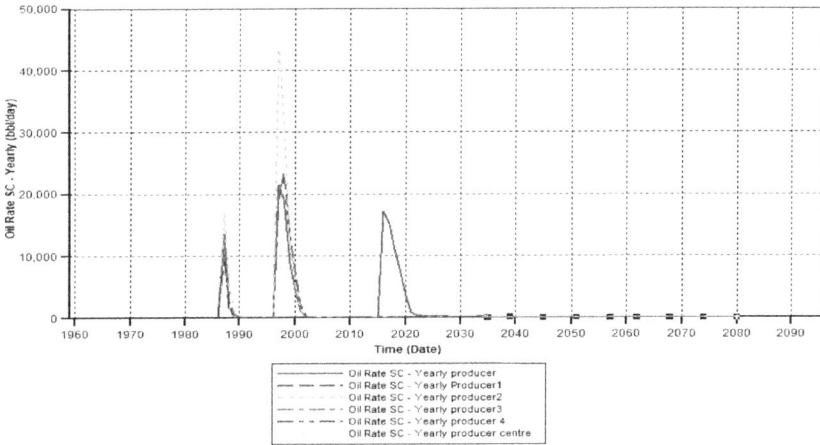

Figure 61: Oil production from wells.

These two diagrams show us that LPG is produced in gaseous form and in liquid form. The gas saturation takes into account the gas phase of the LPG. The oil saturation is composed of residual oil from the reservoir and the liquid phase of the LPG which negates the assumption that the injected gas dissolves in the oil.

4. Balance sheet

These years of oilfield simulation have allowed us to highlight the following facts about underground LPG storage:

-   It is possible to store gas in an exhausted oil field even though there is no free gas initially;
-   This gas is kept in liquid phase until the tank is depressurized;
-   The mixture produced is composed of light oil, water and gas;
-   an increase in the pressure of the tank following a gas injection leads to a slight condensation of the gas;
-   Overpressure of the tank can damage the cover and cause a disaster although it could increase the gas production;
-   it is preferable to produce at a reasonable flow rate in order to maintain the gas in liquid form;
-   abandonment of production can be caused by damage to the tank structure, overproduction of water or a consequent drop in pressure.
-   It is important to limit the production of water as much as possible and to keep the bubble pressure high in order to limit the expansion of the gas phase;
-   the presence of a fault in the reservoir limits the movement of fluids and the ability of surrounding wells to draw and inject;
-   studies on the chemical reactions between the fluid and the gas to be injected must be carried out before the beginning of the injection operations;
-   it is possible to store other types of gas;
-   as injections and withdrawals are made, the conditions of the reservoir and the nature of the fluid change;
-   The implementation of this type of storage requires years of reservoir studies and planning to determine an optimal production scenario;
-   It is preferable to have several injectors for 1 producer in order to maintain the pressure on both sides of the producer well, which will promote fluid migration and limit degassing. It is also desirable to inject and produce from the same well while keeping at least one control well and water injector in case the reservoir needs to be purged;
-   to evacuate the gas phase, it is necessary to inject water.

General conclusion

The aim of this work was to simulate the underground storage of gas in depleted petroleum layers and to highlight the main mechanisms of flow and storage of gas in the reservoir. During this work, we proceeded to the primary and secondary recovery of hydrocarbons and then to the storage and production of gas in this reservoir. This allowed us to determine the optimal conditions to efficiently store gas in the underground.

During the injection tests, we were able to observe a phenomenon of slowing down of the fluid flow speed at the level of the normal fault. This slowing down varies according to the nature of the fluids as it is more visible during the injection of water than gas. Most of the losses observed during the production of gas in the reservoir are due to its change of physical state. The gas changing state during production occupies the surface layers of the reservoir and limits the storage capacity as it goes along. The interaction between the reservoir pressure and the bubble pressure is the main cause of this change of state of the LPG.

Underground storage in porous media offers enormous prospects because it is possible to reconvert abandoned oil formations into storage of natural gas, liquefied petroleum gas, carbon dioxide (CO_2) or methane made from CO_2 for security purposes and to prevent possible energy crises. The large storage capacity of these underground reservoirs and their resistance to mechanical alterations have favored their development.

Perspectives

This study on underground gas storage in depleted reservoirs can be improved by conducting in-depth studies on the following topics

- the interaction between the normal fault and fluid flow (water, oil and gas);
- Determination of gas expansion volume as a function of temperature and pressure variations;
- Determining the storage capacity of a tank according to the nature of the fluid to be stored;
- study on the safety of rehabilitating depleted deposits in underground storage sites;
- environmental impacts of the implementation of an underground storage site;
- modeling and simulation of underground hydrocarbon storage in mined cavities;
- Modeling and simulation of underground hydrocarbon storage in saline cavities;
- simulation of underground gas storage in aquifers;
- interaction between fluid flow and the presence of a reverse fault;
- Historical correlation and simulation of the rehabilitation of a real oil reservoir into a gas storage site;
- planning of surface equipment for each type of underground storage;
- study the stability of the caprock as a function of gas injection into a reservoir.

Bibliography

[1] André CLERC-RENAUD, "PETROLEUM - Storage", Encyclopædia Universalis, 2020.

[2] Ali GAYA, "The underground storage of hydrocarbons: Situation in the world and opportunities for Tunisia", article, accessed on 14/08/20 on the site https://www.leaders.com.tn/article/29789 ,2020.

[3] Quentin SOUBRANNE, "With the world economy in a state of collapse, we will soon no longer know where to store the oil that is flowing", article, BFM Business, 2020.

[4] KUIEKEM Donald, "Optimization of safety for the storage of liquefied petroleum gas (LPG) in a filling center: the case of BOCOM GAS," engineering thesis, School of Geology and Mining (EGEM), 2018.

[5] Hélène GIOUSE, "Les stockages souterrains de gaz naturel," article accessed 8/15/20 at https://www.mines-paris.org/ , 2015.

[6] CMG (Computer Modelling Group) General release, accessed on 20/08/2020 at https://www.cmgl.ca/software ,2020.

[7] Abdelaziz EL-HOSHOUDY and Saad DESOUKY, "PVT Properties of Black Crude Oil," Article, accessed 08/15/20 at https://www.intechopen.com/books/ , doi:10.5772/intechopen.82278,2019.

[8] Susana Jiménez MORALES, "Recuperacion secundaria en campos de petroleo y su conversion en almacenamientos subterraneos de gas natural," PhD thesis, Department of Applied Mathematics, 2012.

[9] Saeid MOKHATAB, John Y. MAK, "Handbook of Natural Gas Transmission and Processing," book, (4th edition, pp.62-24), 2019.

[10] Leonid F.KHILYUK, Bernard ENDRES, "Gas Migration ", article, accessed on 21/08/20 at https://www.sciencedirect.com , 2020.

[1] A.C.LACOSTE and P.BEREST, "Stockages souterrains d'hydrocarbures : sécurité et protection de l'environnement", article, Revue française de Géotechnique, doi : 10.1051/geotech/198114b061, 1981.

[2] A. Lallemand-Barres, " Le stockage souterrain ", rapport, étude documentaire par le Bureau de Recherches Géologiques et Minières, consulted on 24/08/20 at http://infoterre.brgm.fr/rapports/75-SGN-097-AME , 1975.

[3] Bulent OZTURK, "Simulation of depleted gas reservoir for underground gas storage", PhD thesis at the Middle East Technical University (pp. 60-70), 2004.

[4] Vincent ROCHE, "Architecture et croissance des failles dans les alternances argilo-calcaires Exemples dans les alternances du Bassin du Sud-Est (France) et modélisation numérique", Thesis Institut de Radioprotection et de Sûreté Nucléaire (pp. 65-80), 2011.

[5] Pierre BEREST, "Underground storage of gases and hydrocarbons: prospects for the energy transition," article, accessed on 1/09/20 at https://www.encyclopedie-environnement.org/, 2018.

[6] GUANGCHUAN Liang et BINGQIANG Zhang, " The Dynamic Simulation of Underground Gas Storage by Computer ", article, ResearchGate, DOI: 10.1109/ICCIS.2011.289, 2011.

[7] Jerzy STOPA, " Simulation and Practice of the Gas Storage in Low Quality Gas Reservoir ", article, International Gas union, 2012.

[28] E.KHAMEHCHI and F.RASHIDI, "Simulation of Underground Natural Gas Storage in Sarajeh Gas Field, Iran", article, ResearchGate, DOI: 10.2118/106341-MS, 2006.

[9] Alvaro SAINZ-Garcia, " Dynamics of underground gas storage. Insights from numerical models for carbon dioxide and hydrogen ", thèse, Centre pour la

Communication scientifique directe, Id: tel-01703296, version 1 (pp 85-95), 2017.

Frank HEINZE, "UNDERGROUND STORAGE", report of the 22nd World Gas Conference (pp.56-71), 2003.

Emmanuel ATANGANA ELOOUNDOU, "Evolution of gas-water flow in aquifer UGS (A-UGS) from gas-water displacement to cyclic storage operation," article, Université technique de Freiberg, 2011.

Manuel Coronado, Oscar Valdiviezo Mijangos, " A New Scheme to Describe Multiwell Compressible Gas Flow in Reservoirs ", article, Researchgate, DOI: 10.1007/s11242-008-9280-2, 2009.

Marco Thiele, "Streamline Simulation Streamline Simulation", conference report, Researchgate, 2003.

Gian Luigi Chierici, " Principles of Petroleum Reservoir Engineering ", article, Researchgate, DOI: 10.1007/978-3-662-02964-0, 1994.

Larry Lake, "Enhanced Oil Recovery," book, Researchgate, 1989.

Marcela Arteaga-Cardona, Javier Molina, ROGELIO HERNANDEZ, " A Successful Gas Injection Pilot Test in a Mature and Complex Fractured Carbonate Reservoir, Oxiacaque Field, Southern Mexico ", article, Researchgate, DOI: 10.2118/114010-MS, 2008.

Cheng Hao, Ichiro Osako, Michael King, " A Rigorous Compressible Streamline Formulation for Two and Three-Phase Black-Oil Simulation ", article, Researchgate, DOI: 10.2118/96866-PA, 2006.

Ginevra Di Donato, Martin J. Blunt, " Streamline-based dual-porosity simulation of reactive transport and flow in fractured reservoirs ", article, Researchgate, DOI: 10.1029/2003WR002772, 2004.

Andrea Tamburini, Giacomo Falorni, " Satellite Interferometry for Reservoir Monitoring and Management - Applications for CCS, UGS and Geothermal

Exploitation ", rapport de conference, Researchgate, DOI: 10.3997/2214-4609.20144155, 2011.

[30] Adil Sbai, " Double porosity - double permeability model of the near geothermal exploitation field at Bouillante ", rapport technique, Researchgate, DOI: 10.13140/RG.2.2.17023.84641, 2007.

[31] Wolf Tilmann Pfeiffer, Sebastian Bauer, " Hydrogen storage in a heterogeneous sandstone formation: Dimensioning and induced hydraulic effects ", article, Researchgate, DOI: 10.1144/petgeo2016-050, 2017.

[32] Jonathan Ennis-King, Paterson Lincoln, " Role of Convective Mixing in the Long-Term Storage of Carbon Dioxide in Deep Saline Formations ", rapport de conference, Researchgate, DOI: 10.2523/84344-MS, 2003.

[35] Davood Zivar, Sunil Kumara, Jalal Foroozesh, " Underground hydrogen storage: A comprehensive review ", article, Sciencedirect, doi.org/10.1016/j.ijhydene.2020.08.138, 2020.

[36] Leszek Lankof, Radosław Tarkowski, " Assessment of the potential for underground hydrogen storage in bedded salt formation ", article, Sciencedirect, doi.org/10.1016/j.ijhydene.2020.05.024, 2020.

[37] Sunil Kumara, jalal Foroozes, Katriona Edlmann, " A comprehensive review of value-added CO2 sequestration in subsurface saline aquifers ", article, Sciencedirect, https://doi.org/10.1016/j.jngse.2020.103437, 2020.

[38] Donald L. Katz, M. Rasin Tek, " Overview on Underground Storage of Natural Gas ", article, OnePetro, doi.org/10.2118/9390-PA, 1981.

[39] Judy Feder, " Recovering More Than 70% From the Johan Sverdrup Field ", article, OnePetro, doi.org/10.2118/0920-0062-JPT, 2020.

[40] William Bailey, "Technology Focus: Reservoir Simulation," article, OnePetro, doi.org/10.2118/0718-0046-JPT, 2018.

Annexes

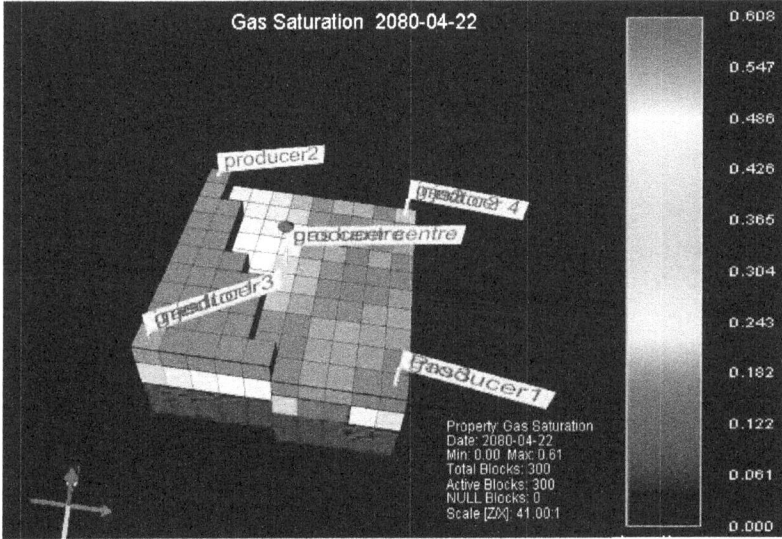

Appendix 1: Gas saturation in 2080.

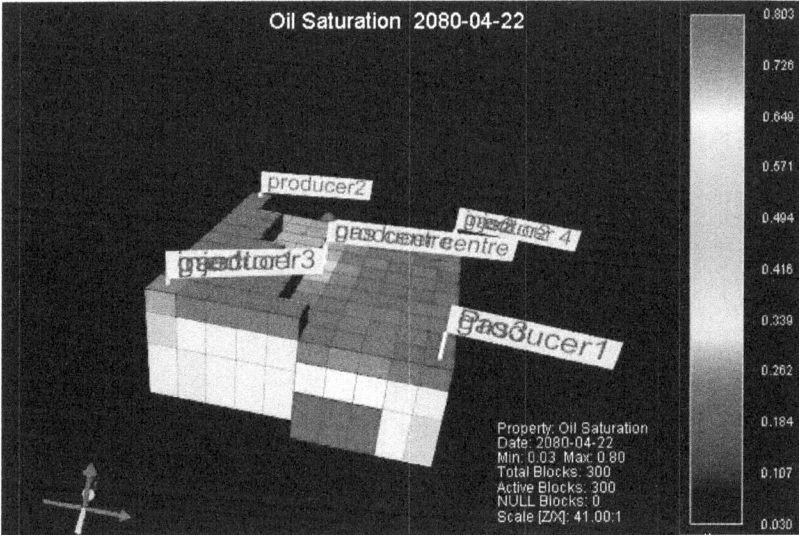

Appendix 2: Oil saturation 2080.

I

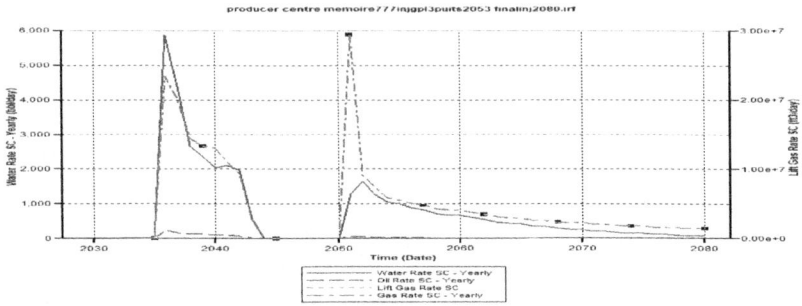

Appendix 3: Monitoring of the central well.

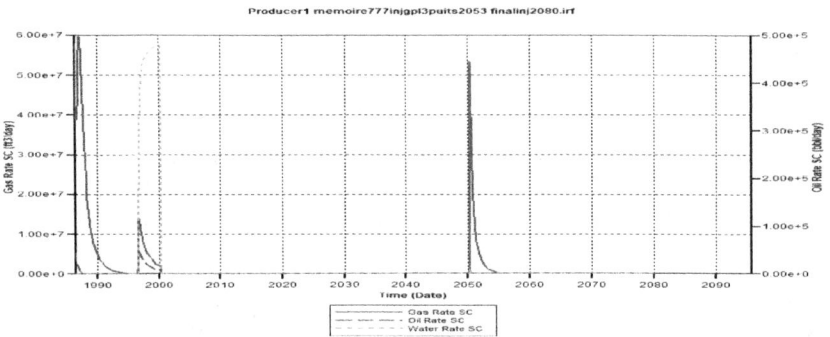

Appendix 4: Monitoring of the 'producer1' well.

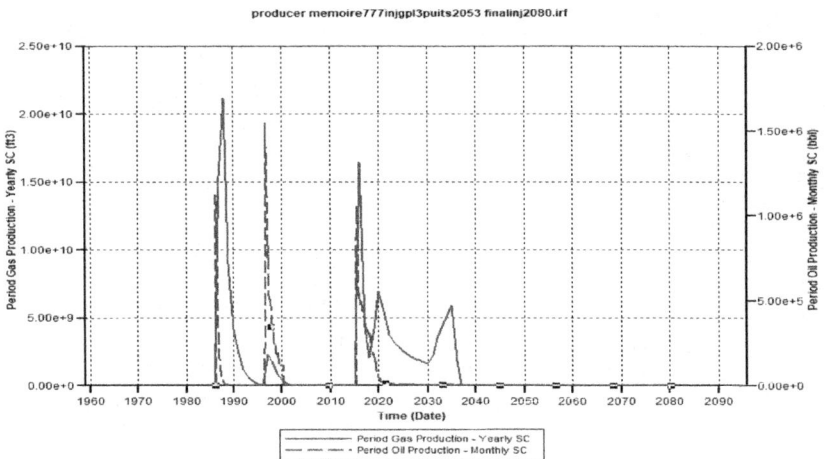

Appendix 5: Monitoring of the 'producer' well.

II

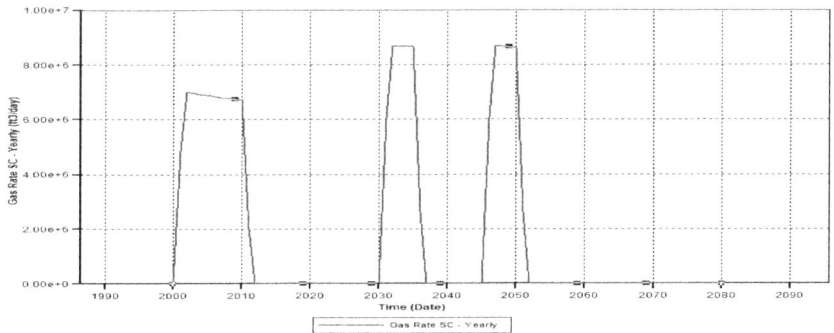

Appendix 6: Monitoring of the 'gas1' well.

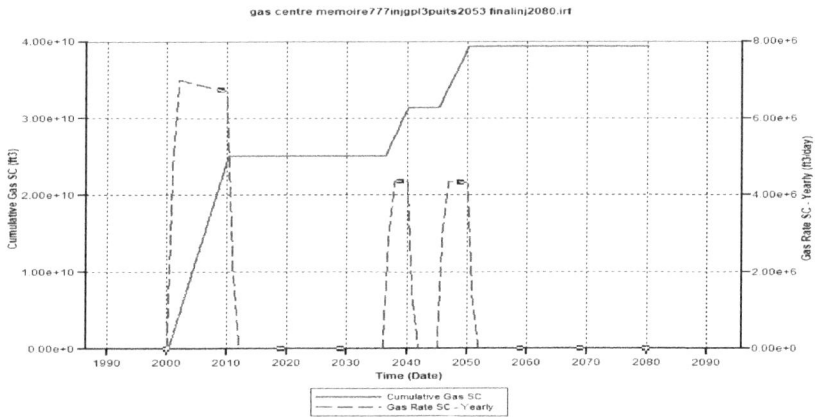

Appendix 7: Monitoring of the gas center well.

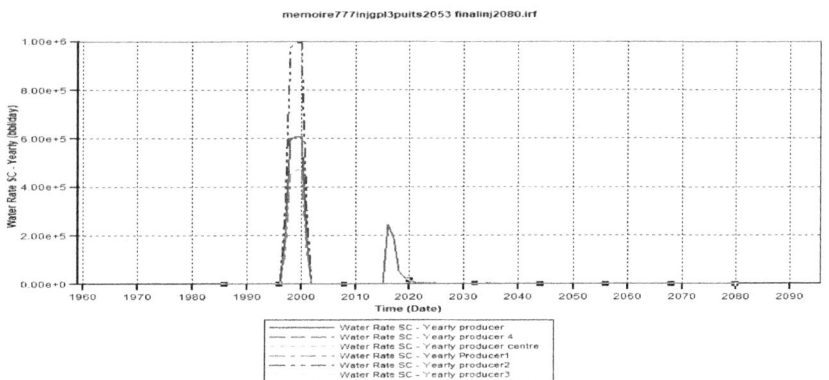

Appendix 8: Water production monitoring.

III

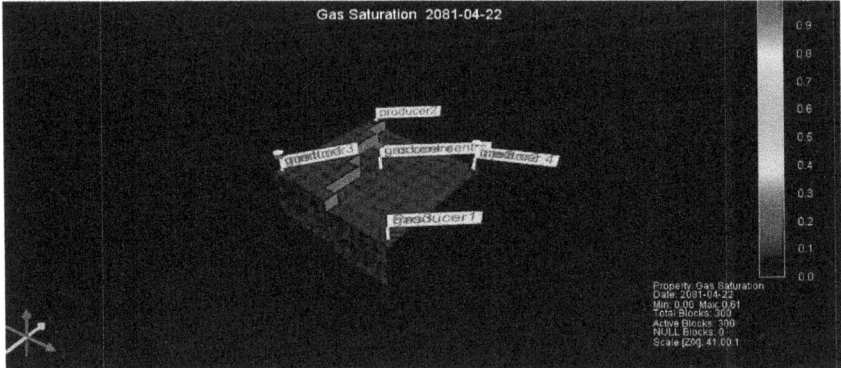

Appendix 9: Gas saturation after water injection in 2081.

```
     Field Total                              Fluid
                          Oil      Gas      Water    Solvent   Polymer
                        -------   -------   -------   -------   -------
                        (MSTB)   (MMSCF)   (MSTB)    (MMSCF)   (MLB)
  Cumulative Production  130174   504973    3412e3      NA       NA
  Cumulative Injection     NA     209361    3915e3      NA       NA
  Cumulative Gas Lift      NA        0        NA        NA       NA
  Cumulative Water Influx  NA       NA       NA         NA       NA
  Current Fluids In Place 142407   13529    606980      NA       NA
  Production Rates          0        0      1840.8      NA       NA
  Injection Rates          NA        0      2017.5      NA       NA

  Timesteps:    569  Newton Cycles:   1870  Cuts:    94  Solver Iterations:
```

Appendix 10: Simulation report after water injection in 2081.

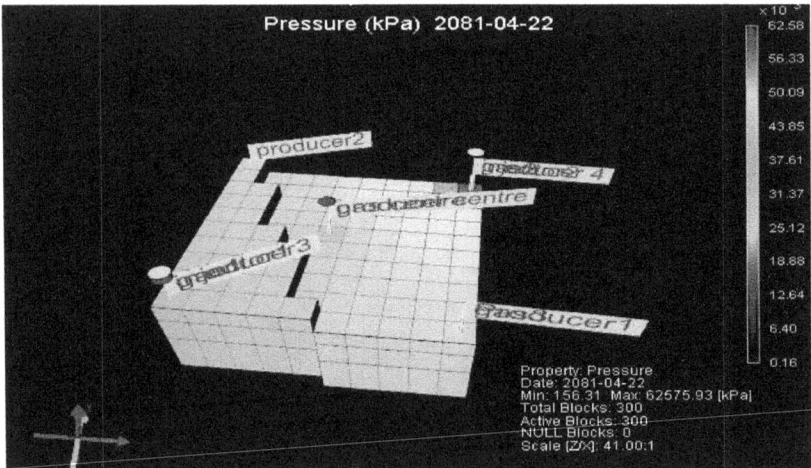

Appendix 11: Pressure distribution in 2081 after injection.

IV

```matlab
function [Y,Xf,Af] = myNeuralNetworkFunction(X,~,~)
%MYNEURALNETWORKFUNCTION neural network simulation function.
%
% Auto-generated by MATLAB, 11-Nov-2020 22:09:38.
%
% [Y] = myNeuralNetworkFunction(X,~,~) takes these arguments:
%
%   X = 1xTS cell, 1 inputs over TS timesteps
%   Each X{1,ts} = Qx5 matrix, input #1 at timestep ts.
%
% and returns:
%   Y = 1xTS cell of 1 outputs over TS timesteps.
%   Each Y{1,ts} = Qx1 matrix, output #1 at timestep ts.
%
% where Q is number of samples (or series) and TS is the number of
% timesteps.

%#ok<*RPMT0>

% ===== NEURAL NETWORK CONSTANTS =====

% Input 1
x1_step1.keep = [1 2 3 5];
x1_step2.xoffset = [1986;20;18;2400];
x1_step2.gain =
[0.142857142857143;0.0333333333333333;0.0350877192982456;0.0008316008316008
32];
x1_step2.ymin = -1;

% Layer 1
b1 = [-
2.4179162614090059513;2.2737853535861316523;1.544093240650137977;1.52978882
26380171872;-
0.10473117879338128544;0.4439222436383900862;0.61815564713722270795;1.3504
399585872526313;1.3730805226470295111;-2.4940448980271145984];
IW1_1 = [1.6695626044623990136 1.7547725643864928724 -
0.4745164904024191655 0.3972906581011604876;-0.81444223102740354037 -
0.63006401425369784608 -2.199285008896999738 0.52493882271951775031;-
2.0179535185823055388 -2.1006819383646702093 1.1743080341043004022 -
0.6881546905938275005;-0.3023478665331466031 3.3156721672099935816 -
0.76264807599492312473 -0.5831585425421010882;1.4118879971867943635
0.9772332174057517058 1.5433666182569047987
0.57898870140933500483;1.6895584246834567388 1.1378353779604486462
0.98395497260029318731 0.7213655673482188108;1.8921791156915481746 -
0.6524360343226077186 -0.28110892318771724119
1.0184836556637160143;0.28820850175189255404 0.90371141030070467615
0.0027041080098493897391 -2.2979554818041649789;0.28326050353969078888
1.3344071658214846643 1.9334287412757769076 2.0609497700787886565;-
1.4165332613933860895 -1.5172237527527394629 1.3773887057040687942 -
0.2177242190987072612];

% Layer 2
b2 = -0.3385172350561792536;
LW2_1 = [-0.26871967760287734928 -0.5607184422390063506 -
0.24452099938467811158 -0.9020717120028164136 -0.875518317213428654 -
0.56478304766190468023 0.58162073208653075618 0.097622349145038075924 -
0.42300422424173239611 -0.06867597754429271921];

% Output 1
y1_step1.ymin = -1;
```

V

```
y1_step1.gain = 0.0384615384615385;
y1_step1.xoffset = 0;

% ===== SIMULATION ========

% Format Input Arguments
isCellX = iscell(X);
if ~isCellX
    X = {X};
end

% Dimensions
TS = size(X,2); % timesteps
if ~isempty(X)
    Q = size(X{1},1); % samples/series
else
    Q = 0;
end

% Allocate Outputs
Y = cell(1,TS);

% Time loop
for ts=1:TS

  % Input 1
  X{1,ts} = X{1,ts}';
  temp = removeconstantrows_apply(X{1,ts},x1_step1);
      Xp1 = mapminmax_apply(temp,x1_step2);

  % Layer 1
      a1 = tansig_apply(repmat(b1,1,Q) + IW1_1*Xp1);

  % Layer 2
      a2 = repmat(b2,1,Q) + LW2_1*a1;

  % Output 1
  Y{1,ts} = mapminmax_reverse(a2,y1_step1);
  Y{1,ts} = Y{1,ts}';
end

% Final Delay States
Xf = cell(1,0);
Af = cell(2,0);

% Format Output Arguments
if ~isCellX
    Y = cell2mat(Y);
end
end

% ===== MODULE FUNCTIONS ========

% Map Minimum and Maximum Input Processing Function
function y = mapminmax_apply(x,settings)
y = bsxfun(@minus,x,settings.xoffset);
y = bsxfun(@times,y,settings.gain);
y = bsxfun(@plus,y,settings.ymin);
end
```

```
% Remove Constants Input Processing Function
function y = removeconstantrows_apply(x,settings)
y = x(settings.keep,:);
end

% Sigmoid Symmetric Transfer Function
function a = tansig_apply(n,~)
a = 2 ./ (1 + exp(-2*n)) - 1;
end

% Map Minimum and Maximum Output Reverse-Processing Function
function x = mapminmax_reverse(y,settings)
x = bsxfun(@minus,y,settings.ymin);
x = bsxfun(@rdivide,x,settings.gain);
x = bsxfun(@plus,x,settings.xoffset);
end
```

Appendix 12: Algorithm.